WILDLAND FIRES AND FIRE MANAGEMENT

WILDLAND FIRES AND FIRE MANAGEMENT

UNITED STATES GOVERNMENT
ACCOUNTABILITY OFFICE

Nova Science Publishers, Inc.
New York

For permission to use material from this book please contact us:
Telephone 631-231-7269; Fax 631-231-8175
Web Site: http://www.novapublishers.com

Library of Congress Cataloging-in-Publication Data

Wildland fires & fire management / GAO.
p. cm.
ISBN 978-1-60692-076-3 (softcover)
1. Wildfires--United States--Prevention and control. 2. Wildfires--United States--Management. I. United States. Government Accountability Office. II. Title: Wildland fires and fire management.
SD421.3.W5546 2008
363.37'9--dc22

2008034456

Published by Nova Science Publishers, Inc. ; New York

CONTENTS

PREFACE

The Forest Service and Interior agencies have initiated a number of steps to address key operational areas previously identified as needing improvement to help federal agencies contain wildland fire costs, but the effects on containing costs are unknown, in part because many of these steps are not yet complete. First, federal firefighting agencies are developing a system to help them better identify and set priorities for lands needing treatment to reduce fuels, but they have yet to decide how they will keep data in the system current. Second, federal agencies have taken some steps to improve how they acquire and use personnel, equipment, and other firefighting assets—such as implementing a computerized system to more efficiently dispatch and track available firefighting assets—but have not yet completed the more fundamental step of determining the appropriate type and quantity of firefighting assets needed for the fire season. Third, the agencies have clarified certain policies and are improving analytical tools that assist officials in identifying and implementing an appropriate response to a given fire, but several other policies limit the agencies' use of less aggressive firefighting strategies, which typically cost less. Fourth, federal agencies, working with nonfederal entities, have recently taken steps to clarify guidance to better ensure that firefighting costs are shared consistently for fires that threaten both federal and nonfederal lands and resources, but it is unclear how the agencies will ensure that this guidance is followed. This new book is an edited and excerpted version of several GAO reports.

Chapter 1

WILDLAND FIRE MANAGEMENT: BETTER INFORMATION AND A SYSTEMATIC PROCESS COULD IMPROVE APPROACH TO ALLOCATING FUEL REDUCTION FUNDS AND SELECTING PROJECTS *

WHY GAO DID THIS STUDY

Recognizing that millions of acres are at risk from wildland fire, the federal government expends substantial resources on thinning brush, trees, and other potentially hazardous fuels to reduce the fire risk to communities and the environment. However, questions have been raised about how the agencies responsible for wildland fire management—the Department of Agriculture's Forest Service and the Department of the Interior's (Interior) Bureau of Indian Affairs (BIA), Bureau of Land Management (BLM), Fish and Wildlife Service (FWS), and National Park Service (NPS)—allocate their fuel reduction budgets and select projects.

GAO was asked to report on the agencies' processes for allocating funds and selecting projects, and on how, if at all, these processes could be improved to better ensure that they contribute to the agencies' overall goal of reducing risk. To obtain this information, GAO visited headquarters and field offices of all five agencies; obtained data on fuel reduction funding and accomplishments; and reviewed previous evaluations of the fuel reduction program.

ABBREVIATIONS

BIA	Bureau of Indian Affairs
BLM	Bureau of Land Management
FPA	Fire Program Analysis
FWS	Fish and Wildlife Service
GAO	Government Accountability Office
GIS	Geographic information system
HFI	Healthy Forests Initiative

* This Chapter is an excerpted indexed edition of GAO Report GAO-07-1168, Dated September 2007

HFRA	Healthy Forests Restoration Act
NEPA	National Environmental Policy Act
NPS	National Park Service
USDA	United States Department of Agriculture
WUI	Wildland-urban interface

September 28, 2007

The Honorable Norman D. Dicks -Chairman
The Honorable Todd Tiahrt - Ranking Member
Subcommittee on Interior, Environment, and Related Agencies
Committee on Appropriations
House of Representatives

The Honorable Raúl M. Grijalva - Chairman
The Honorable Rob Bishop - Ranking Member
Subcommittee on National Parks, Forests, and Public Lands
Committee on Natural Resources
House of Representatives

The Honorable Greg Walden- House of Representatives

Decades of fire suppression in the practices as logging followed by dense tree planting, have resulted in the accumulation of brush, small trees, and other vegetation that can fuel wildland fires. Similarly, the decades of fire suppression and livestock overgrazing, have degraded ecosystems and made the rangelands vulnerable to the invasion of flammable, nonnative species, such as cheat grass. This accumulation and alteration of vegetation, as well as drought and other stresses related to climate change, have fueled wildland fires. Collectively, these fires have cost billions of dollars to suppress, forced thousands from their homes, and damaged cultural and natural resources. The impacts of these fires have intensified as more and more communities develop in areas that are adjacent to fire-prone lands—the wildland-urban interface.

In response to the increasing threat of wildland fires, the federal agencies responsible for wildland fire management developed the National Fire Plan.[1] These agencies are the Department of Agriculture's Forest Service and the Department of the Interior's (BIA), Bureau of Land Management (BLM), Fish and Wildlife Service (FWS), and National Park Service (NPS). Two components of the National Fire Plan are a 10-year strategy and an implementation plan for protecting communities and the environment that were developed in 2001 and 2002, respectively, by the Secretaries of Agriculture and of the Interior, along with governors of western states and other interested parties, and updated in December 2006. The 2002 plan emphasized reducing hazardous fuel in forests and rangelands to mitigate the risk from wildland fire. In 2003, Congress passed the Healthy Forests Restoration Act (HFRA),[2] with the stated purpose of reducing wildland fire risk to communities, municipal water supplies, and other at-risk federal land through a collaborative process of planning, setting priorities, and implementing fuel reduction projects.[3] HFRA also authorized grants to commercial facilities that use biomass—that is, small-diameter trees and branches—to offset the costs incurred to purchase biomass. Fuel reduction projects can generate substantial amounts of biomass.

According to the updated 10-Year Strategy Implementation Plan, the goal of the fuel reduction program is to reduce the risk of wildland fire to communities and the environment. Fuel reduction projects—using prescribed fire, mechanical thinning, herbicides, grazing, or combinations of these methods—are intended to remove or modify wildland fuel to reduce the potential for severe wildland fires, lessen the damage caused by fires, limit the spread of flammable invasive species, and restore and maintain healthy ecosystems. Local land management units, such as national forests and parks, are typically responsible for selecting and implementing fuel reduction projects.

Since 2001, Congress has appropriated more than $3.2 billion in fuel reduction funds to the Forest Service and Interior. For 2007, the Forest Service received about $300 million, and Interior received about $200 million.[4] After receiving its annual appropriation, the Forest Service allocates funds to its nine regional offices which in turn allocate funds to individual national forests and grasslands. Interior, upon receiving its annual appropriation, allocates funds to BIA, BLM, FWS, and NPS. BLM generally receives about 50 percent of Interior's funding, BIA about 20 percent, and FWS and NPS about 15 percent each. These agencies then allocate funds to their regional or state offices, which, in turn, allocate funds to individual field units, such as national parks or wildlife refuges. (BIA, FWS, and NPS have regional offices, while BLM has state offices. For the purposes of this report, we refer to all of these as regional offices when we discuss the Interior agencies collectively.) Figure 1 shows the annual appropriation and allocation process.

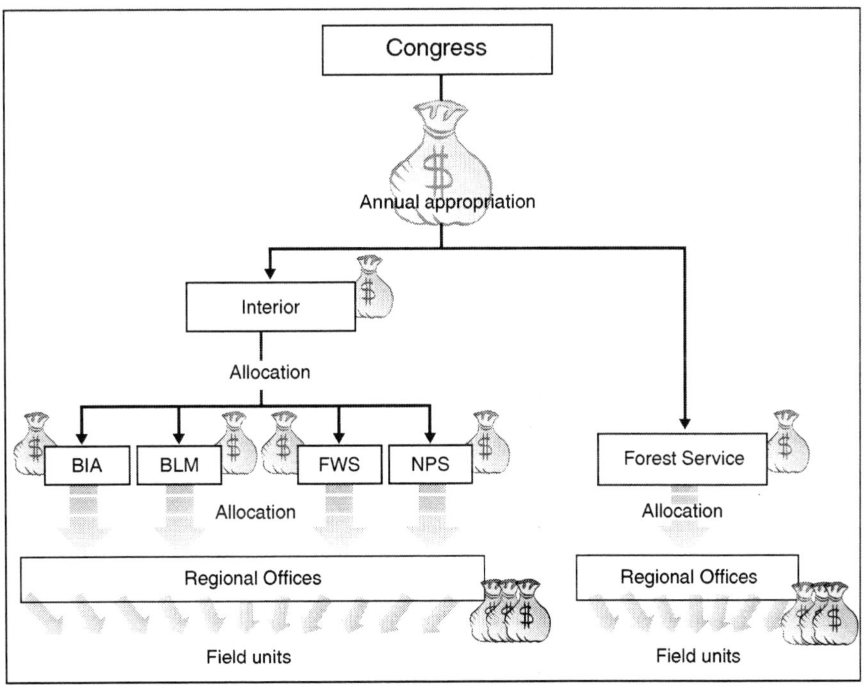

Source: GAO (data) and Art Explosion (clipart).

Figure 1. Annual Appropriation and Allocation Process for Fuel Reduction Funds.

Recognizing that treating all of the land in need of fuel reduction may take decades, the agencies have acknowledged the importance of setting priorities for which lands are to receive treatment so that they can select those treatments that will be the most effective at reducing the risks from wildland fire. However, we have found a long-standing pattern of shortcomings in the processes the Forest Service and Interior agencies use to identify and set priorities for lands needing fuel reduction. Between 1999 and 2003, we reported that the Forest Service and Interior had made it a priority to treat lands at the highest risk from wildland fire, but they had not identified the amount or location of such lands and had not issued guidance specific enough for field staff to set priorities for individual projects. We concluded that the agencies needed a cohesive strategy outlining long-term options and associated costs for reducing fuel.[5] In subsequent reports, we noted, among other things, the progress the agencies had made in improving their data, but reiterated that they needed to complete ongoing efforts to identify lands at risk from wildland fire—by collecting information on the hazards, the likelihood of fire occurring, and the values at risk—so funds could be targeted to such lands. We also reiterated the need to develop a cohesive strategy that included long-term options and associated costs so that Congress could make informed decisions about cost-effective approaches to fuel reduction.[6]

In this context, you asked us to report on the agencies' current processes for identifying and setting priorities for fuel reduction. Specifically, you asked us to (1) identify the processes the Forest Service, Interior, and the four Interior agencies use to allocate fuel reduction funds and select projects for implementation, including the factors that influence these processes, and (2) determine how, if at all, these processes could be improved to better ensure that they contribute to the agencies goal of effectively reducing the risk of wildland fire to communities and the environment.

To address these objectives, we met with national, regional, state, and local officials of the Forest Service, Interior, and Interior agencies. At the national level, we met with agency officials at their Washington, D.C., headquarters, as well as at the National Interagency Fire Center in Boise, Idaho. At the regional and state levels, we used a structured interview guide to speak, in person or by telephone, with officials in all Forest Service regional and BLM state offices, as well as with officials in selected BIA, FWS, and NPS regional offices that collectively received a substantial portion of each agency's fuel reduction funds. At the local level, we visited 20 local units, such as national forests and BLM field offices, in eight states to gain a better understanding of their processes for selecting fuel reduction projects for implementation. We selected local units that are diverse in geographic location, predominant vegetation type, and proximity to communities and development. We also obtained and reviewed applicable laws, regulations, and agencywide and regional policies; agency data on funding allocations; and electronic data on the extent of agency fuel treatment activities. We tested these data and found that they were sufficiently reliable for the purposes of this review. Finally, we interviewed several nonfederal parties, including representatives from environmental groups and the Western Governors' Association.[7] We conducted our work from August 2006 to September 2007 in accordance with generally accepted government auditing standards. See appendix I for a detailed description of our methodology.

RESULTS IN BRIEF

In allocating fuel reduction funds and selecting projects, the Forest Service—at the national, regional, and local levels—uses both quantitative processes (such as computer models or scoring systems) and professional judgment and, in doing so, considers multiple factors, such as risk assessments, treatment cost per acre, and collaboration with communities or other entities. Specifically, for 2007, we found the following:

- *At headquarters,* the Forest Service began using a computer model to influence funding allocations to regions. To set priorities for each region's fuel including some intended to assess risk, such as the potential for fires occurring in each region and their expected severity, as well as other factors, such as regional use of biomass removed in fuel treatments and treatment cost per acre. However, the Forest Service's funding allocations to its regions were not consistent in all cases with the priority scores resulting from the model, with some high-scoring regions receiving less funding than some lower-scoring regions. These disparities occurred for a number of reasons, such as the higher costs of fuel reduction in some areas. However, the model did not substantially influence the agencys' 2007 allocations, instead, the Service relied largely on prior year funding levels and used results from the model only to make minor adjustments. Officials said they expect the model's results decisions, but curbed this influence initially because they were still refining the model and wanted to maintain relatively stable regional funding levels.
- *In the regions,* each region determined how to allocate funds to national forests and what factors to consider, as long as they were consistent with the factors used in the national allocation model. Four of the Service's their allocation processes, while five relied primarily on professional judgment. For example, the Rocky Mountain region used a computer model that evaluated data on multiple factors, such as vegetative conditions and areas of insect-killed trees, while the Southern region convened a group of officials who used professional judgment and considered historical funding levels, the capabilities of the forests, per-acre treatment cost, local priorities, and acreage targets when allocating funds. Beginning with the 2008 allocations, the Forest Service plans to require regions to use the headquarters model to inform allocation decisions.
- *Locally,* national forests had discretion in determining how to select projects. Some used quantitative, data-driven processes, while others relied primarily on professional judgment or collaborative processes involving other agencies and local communities. Forests considered a range of factors—similar to those used at the national and regional levels—when selecting projects.

Like the Forest Service, Interior and its agencies' national, regional, and local offices used both quantitative and judgmental processes for allocating fuel reduction funds and selecting projects and considered multiple factors that are similar to those the Forest Service uses. More specifically, for fiscal year 2007, we found the following:

- *Interior* allocated funds to its four agencies primarily on the basis of historical funding levels; however, Interior is developing a computer model similar to the Forest Service's and it used the model to allocate 5 percent of its funds in 2007. Interior agencies, in turn, had the flexibility to determine how to allocate funds, within the parameters of departmental guidance.
- *BLM* headquarters allocated funds primarily on the basis of historical funding levels, but officials told us that, starting in 2008, they plan to use a quantitative process incorporating multiple factors, such as the potential for fires to occur, treatment cost, and local risk ratings.
- *BIA* headquarters allocated funds using quantitative processes, but these processes generally emphasized a single factor—BIA units' past performance (measured in acres treated) in carrying out fuel reduction activities.
- *FWS* headquarters allocated funds to regional offices using a quantitative process—a model that considers a range of factors, such as the history of fires, fuel conditions, and communities at risk.
- *NPS* headquarters allocated funds to regions largely on the basis of historical funding levels. These funding levels were originally determined using a model that assessed the risk from wildland fire, among other factors.

At the regional and local levels, regional offices and field units in all four agencies used a variety of processes to allocate funds and select projects for implementation. Some processes emphasized quantitative data, while others emphasized professional judgment. The regional and local offices also considered a range of factors, consistent with departmental direction.

Although the Forest Service and Interior have begun taking action to enhance their funding allocation processes, there are additional steps they could take to improve these processes to better ensure they advance the agencies' goal of communities and the environment. Specifically, the agencies could improve their processes by taking the following five steps:

- *Consistently using risk assessments.* The agencies did not consistently use risk assessments in their 2007 allocation processes at the national, regional, and local levels, in some cases because national or regional offices expected local units to do so. However, agency officials cannot be sure that projects identified as high risk locally would likewise be the highest risk from a regional or national perspective. Even when the agencies did conduct risk assessments, they found it difficult to meaningfully distinguish between higher- and lower-priority locations because one key value at risk—the wildland-urban interface—is broadly defined and many different areas are classified as interface. Although the agencies' guidance sets a priority on projects in the interface, it does not specify whether some of the areas classified as interface ought to be higher priority than others. As a result, projects as diverse as those protecting remote power lines, individual ranch houses, or large suburban subdivisions can all fall within the wildland-urban interface category and, thus, be designated high priority—complicating agency officials' attempts to identify and direct their limited resources toward the highest-priority areas.

- *Developing information on cost effectiveness.* The agencies often considered costs when allocating funds, but not cost effectiveness— primarily because they lack information on treatment effectiveness. Without such information, it is difficult to know whether a treatment's cost is warranted or to compare the cost effectiveness of different potential treatments to decide how to optimally allocate funds.
- *Developing information on cost effectiveness.* The agencies often considered costs when allocating funds, but not cost effectiveness— primarily because they lack information on treatment effectiveness. Without such information, it is difficult to know whether a cost is warranted or to compare the cost effectiveness of different potential treatments to decide how to optimally allocate funds.
- *Clarifying the importance of factors unrelated to risk or effectiveness.* The agencies often considered factors other than risk, treatment effectiveness, and cost effectiveness when allocating funds and selecting projects. When these external factors—such as funding stability and the use of biomass resulting from fuel reduction treatments—have considerable influence, it is difficult for the agencies to ensure that they are allocating funds so that treatments will most effectively reduce risk.
- *Applying more systematic processes.* The agencies sometimes relied exclusively on professional judgment when allocating funds or selecting projects. Although judgmental processes might result in allocations that maximize risk reduction, the agencies cannot be assured that they routinely do because such processes are not necessarily systematic—that is, methodical, based on criteria, and applied consistently.

To improve the agencies' directed to most effectively reduce the risk from wildland fire, we are recommending that the Secretaries of Agriculture and of the Interior take actions to implement a more systematic allocation process; develop additional information on risk, treatment effectiveness, and cost-effectiveness to support the process; and clarify the relative importance of multiple criteria for setting priorities in allocation and project selection decisions. We provided a draft of this report to the Secretaries of Agriculture and of the Interior for review and comment. The Forest Service and the Department of the Interior generally agreed with our report; their joint comment letter is presented in appendix IV.

BACKGROUND

Wildland Fire Is a Natural Process

Although its effect on communities can be devastating, wildland fire is a natural and necessary process that provides many benefits to ecosystems, such as maintaining habitat diversity, recycling soil nutrients, limiting the spread of insects and disease, and promoting new growth by causing the seeds of fire-dependent species to germinate. Wildland fire also periodically removes brush, small trees, and other vegetation that can otherwise accumulate and increase the size, intensity, and duration of subsequent fires. However, human uses and land management practices— including decades of wildland fire suppression—have excluded fire from ecosystems, reducing the normal frequency of wildland fire and subsequently causing an accumulation of vegetation. Federal researchers have estimated that unnaturally dense fuel accumulations on 90 million to 200 million acres of federal lands in the contiguous United States place these lands at an elevated risk of severe wildland fire and that these conditions also hold true for many nonfederal lands.

Most lands in the United States evolved with fire, and each ecosystem has a characteristic fire regime that describes the role fire plays in the ecosystem, including typical fire frequency, scale, intensity, and duration. These regimes are numbered I through V, with fire regime I characterized by low-severity fires that historically occurred every 35 years or less, fire regime II characterized by high-severity fires that historically occurred every 35 years or less, fire regime III characterized by mixed-severity fires that historically occurred every 35 to 100 or more years, fire regime IV characterized by high-severity fires that historically occurred every 35 to 100 or more years, and fire regime V characterized by high-severity fires that historically occurred every 200 or more years. Many ecosystems— particularly those in fire regimes I and II—have now missed numerous fire cycles as a result of past suppression policies and other land management practices. This departure from the natural fire regime is categorized by a measure called condition class, which the agencies have used as a generalized rating for the risk of uncharacteristic wildland fires that may cause undesirable ecological consequences. Ecosystems in condition class 1 are generally within their historical fire return interval, so fires in these areas pose little risk to natural processes—although fires in such ecosystems may still pose a high risk to communities. Areas in condition classes 2 and 3 have moderate to significant departures from historical fire experiences. In such areas, fuel that would typically have burned periodically has instead accumulated, posing a higher risk that uncharacteristically large amounts of vegetation and other natural resources would be lost from wildland fire; fires in these areas may also pose a high risk to communities.

Five Agencies Are Responsible for Wildland Fire Management

The Forest Service, BIA, BLM, FWS, and NPS are responsible for wildland fire management, including fuel reduction. These five agencies manage about 700 million acres of land in the United States, including national forests, national grasslands, Indian reservations, national parks, and national wildlife refuges. The Forest Service and BLM manage the majority of these lands, with the Forest Service managing about 190 million acres and BLM managing about 260 million acres; BIA, FWS, and NPS each manage less than 100 million acres. Figure 2 shows the distribution of land among the five agencies. Each agency has between 7 and 12 regional offices that oversee field units.

Each year, the Forest Service, Interior, and Interior agencies set performance targets for region- and state-level fuel reduction by establishing the number of acres the agencies expect to be treated—both within and outside of the wildland-urban interface—using the funds allocated. For example, for fiscal year 2007, BLM assigned a target of almost 90,000 acres to its Oregon/Washington state office and specified that two-thirds of the acres should be in the wildland-urban interface. Between 2001 and August 2007, land managers treated more than 18 million acres under the fuel reduction program, including about 8.5 million acres near communities. [8] These acres include federal, state, and private land, because, in addition to conducting fuel treatments on federal lands, the agencies work with and grant funds to local communities to conduct fuel reduction treatments on state and private lands. These acres also include those that have been treated more than once.

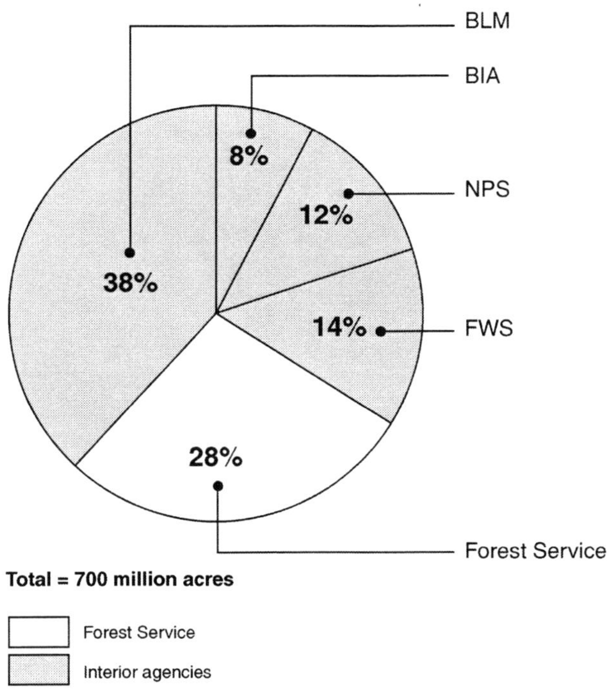

Source: GAO analysis of Forest Service and Interior data.

Figure 2. Distribution of Total Land Managed by the Forest Service, BIA, BLM, FWS, and NPS.

The agencies generally reduce fuel using either mechanical treatments, in which equipment—such as chainsaws, chippers, bulldozers, or mowers— is used to cut vegetation, or prescribed burning, in which fires are deliberately set by land managers to restore or maintain desired vegetation conditions.[9] Figure 3 depicts a mechanical thinning project, and figure 4 depicts a prescribed burn.

Although prescribed burning can be risky, burning under specified fuel and weather conditions enables fire to be controlled at a relatively low intensity level within a confined area. Prescribed burning is very effective in removing smaller vegetation, such as grasses, leaves, pine needles, and twigs, but is not as effective in removing larger fuel, such as trees, or in thinning stands to desired densities. In contrast, mechanical treatment methods are effective in thinning stands and removing larger vegetation but may increase the amount of smaller fuel on the ground, including tree tops and limbs (referred to as slash) and other debris from thinning. As a result, some fuel reduction projects use multiple treatment methods and may span several years. For example, a field unit may first treat an area mechanically to thin accumulated vegetation and then follow with a prescribed burn to remove remaining slash and litter on the ground.

Source: BLM.

Figure 3. A Mechanical Thinning Project for Fuel Reduction on BLM Land in California.

Source: Forest Service.

Figure 4. Prescribed Fire for Fuel Reduction on Forest Service Land in South Carolina.

In addition to reducing the risk of fire to communities and the environment, one of the long-term goals of the fuel reduction program is to allow fire to resume its natural role. By conducting treatments, including creating fire breaks to help contain the spread of fire, the agencies increase the amount of land where naturally ignited fires can safely be allowed to burn. Under wildland fire use policies, land managers may allow wildland fires that are naturally ignited to continue to burn, as long as fuel and weather conditions are appropriate and the fire is located within an area designated for wildland fire use.[10] Managers are thus able to use natural fire to meet resource objectives, such as removing excess vegetation.

Although the five agencies all reduce fuel in order to reduce risk to communities and the environment, their fuel reduction programs reflect differences in their missions, predominant vegetation types, and allowable land uses. For example, FWS's wildlife habitat, and the agency generally conducts more prescribed burns than mechanical treatments because such burns frequently improve habitat as well as reduce risk; the agency has been conducting prescribed burns since the 1930s. Similarly, prescribed burns, as well as wildland fire use, are the preferred fuel treatment methods at NPS and have been used by the agency for decades. NPS prefers these treatment methods over mechanical treatments because its mission emphasizes preservation of natural and cultural resources, and fire is a natural process that better aligns with this mission. Regarding predominant vegetation types, BLM's lands are largely rangelands, while lands managed by other agencies, such as the Forest Service and FWS, include more forests. As a result of this difference, BLM not only conducts mechanical treatments and prescribed burns, as do the other agencies, but also uses herbicides to reduce fuel, especially where rangelands have been invaded by exotic plants such as cheat grass. Agency differences in allowable land use also affect their fuel reduction programs. For example, the Forest Service, BIA, and BLM have active commercial timber programs, and field units may therefore conduct fuel treatments that benefit both the timber and fuel reduction programs. BIA and NPS also manage lands with numerous archaeological sites, which must be considered when conducting treatments. In contrast, the majority of BLM's land coordinates fuel treatments with potentially affected ranchers.

Expansion into the Wildland-Urban Interface Has Increased, as Has the Federal Focus on Wildland Fire Management

Urban and suburban expansion into the wildland-urban interface has increased the number of communities and structures at risk of wildland fire near federal lands that the five agencies manage. Experts estimate that almost 60 percent of all new housing units built in the 1990s were located in the wildland-urban interface and that this growth trend continues. They also estimate that more than 30 percent of housing units overall are located in the wildland-urban interface and that the interface covers about 9 percent of the nation's infrastructure are located in the wildland-urban interface, including power lines, campgrounds and other recreation facilities, oil and gas wells, communications towers, and roads.

After the National Fire Plan was developed, the agencies began receiving sharp increases in funding for fuel reduction and, since 2001, Congress has appropriated between about $400 million and $500 million annually for fuel reduction under the plan. (App. II shows agency fuel reduction funding appropriations and allocations for 2005 through 2007; app. III shows the agencies'accomplishments.) In 2002, the President announced the Healthy Forests Initiative (HFI), directing the departments of Agriculture and of the Interior and the Council on Environmental Quality to provide regulations to ensure more timely decisions, increase efficiency, and improve results in reducing the risk of catastrophic wildland fires.

In 2003, Congress passed HFRA to reduce wildland fire risk to communities, municipal water supplies, and other at-risk federal lands through a collaborative process of planning, setting priorities for, and implementing fuel reduction projects. In funding authorized fuel reduction projects on federal land, HFRA requires the agencies to use at least 50 percent of these funds in the wildland-urban interface.[11] The act also established separate environmental analysis and administrative review procedures for fuel reduction projects authorized under HFRA. In providing assistance for fuel reduction activities on nonfederal lands, HFRA requires the agencies, to the maximum extent practicable, to give priority to communities that have adopted a community wildfire protection plan (community plan) or have taken proactive measures to encourage willing property owners to reduce fire risk on private property. A community plan identifies and sets priorities for fuel reduction treatments and recommends the types and methods of treatment on federal and nonfederal land that will protect at-risk communities and essential infrastructure; community plans also recommend measures to reduce structural ignitability throughout the at-risk community. These plans are to be agreed upon by the applicable local government, local fire department, and state forest management agency, in consultation with other interested parties and the federal land management agencies. As of February 2007, there were at least 1,100 completed community plans covering almost 3,300 communities throughout the United States, and approximately 450 additional plans in progress, according to the National Association of State Foresters. A community plan may cover one or more communities, and some cover entire counties.

According to the 10-Year Strategy Implementation Plan, the goal of the fuel reduction program is to reduce the risk of wildland fire to communities and the environment. However, some fuel treatments provide other benefits in addition to this overall program goal; for example, agency staff sometimes conduct prescribed burns to both reduce fuel and enhance wildlife habitat, or conduct mechanical thinning projects before a commercial timber sale. Similarly, in addition to the approximately $400 to $500 million appropriated for fuel reduction each year, funds from other agency programs, such as wildlife management or timber, often are used to conduct vegetation treatment projects that reduce fuels as a secondary benefit. In addition, the agencies sometimes receive partnership funding from outside organizations, such as the Rocky Mountain Elk Foundation or The Nature Conservancy, to conduct collaborative treatments.[12]

GAO Has Reviewed Agencies' Fuel Treatment Programs

While the federal agencies acknowledge the importance of setting priorities for lands needing fuel treatments, we have identified a long- standing pattern of shortcomings in the processes the Forest Service and Interior use to identify and set priorities for these lands. Between 1999 and 2007, we conducted several reviews of the agencies' wildland fire management efforts, including the fuel reduction program. We found that while the agencies aimed to target fuel reduction efforts to the highest risk areas, they could not ensure that they were doing so. For example, in 1999, we found that the Forest Service intended to give priority to treatments in the wildland-urban interface but was hampered in doing so because it had not fully defined and mapped such areas.[13] We concluded that the Forest Service needed a cohesive strategy outlining options and associated costs for reducing fuel. We reiterated the agencies' need for a cohesive strategy in several additional reports and testimonies issued between 2002 and 2007.[14] In 2000 and 2002, we reported that the Forest Service and Interior did not know how many communities were at high risk of severe wildland fire or their locations and the cost to treat them and, therefore, could not set treatment priorities.[15] We further reported in 2002 and 2003 that the agencies did not have quantifiable long-term and annual performance measures to assess progress in reducing the risks of wildland fire and that they measured the performance of the fuel reduction program by number of acres treated, which does not necessarily correlate to risk reduction.[16]

Similarly, in 2004, we reported that the agencies did not systematically assess the risks to environmental resources and ecosystems and, therefore, could not target fuel reduction efforts to the resources and ecosystems at highest risk. To set priorities for fuel reduction activities, the agencies must first identify areas at risk from wildland fire by considering three elements: hazard, risk, and value.[17] A hazard is a potential event, such as a wildland fire, and the conditions that cause it; in the case of wildland fire, both the fuel conditions and the fire itself are the hazard. Risk is the probability that an event such as a wildland fire will occur. Values are the resources and property that could be lost or damaged because of a hazard; in the case of wildland fire, values might include social, economic, or environmental values.[18] Without considering all three elements, the agencies may not be appropriately setting priorities for areas needing fuel reduction. For example, an area with high vegetation hazard may not be in an area where fires are likely to occur, making it a lower priority for treatment; likewise, a high hazard area might not be near something of value that could be lost or damaged in a fire, also making it a lower priority for treatment.

We also found, through multiple reviews, that the agencies could benefit from coordinating their efforts to manage wildland fires because wildland fire is a shared problem that transcends administrative boundaries. For example, in 2001 we reported that federal policy for managing wildland fire required coordination, consistency, and agreement among the Forest Service, Interior, and Interior agencies, but we found that the agencies planned and managed wildland fire management activities largely on agency-by-agency and unit-by-unit bases, and could not ensure, among other things, that they were allocating funds to the highest-risk communities and ecosystems.[19] In a 2002 report, we noted that the Forest Service and Interior had either developed or were in the process of developing numerous strategies that had different goals and objectives and that were not linked, primarily because the agencies had been managing their lands on an agency-by-agency basis for decades.[20] In a subsequent testimony, which emphasized the agencies' need for a cohesive strategy as well as clearly defined and effective leadership, we concluded that effectively addressing wildland fire would require a sustained and coordinated effort between departments.[21] (See Related GAO Products.)

The Forest Service Uses a Mix of Quantitative and Judgmental Processes and Considers a Range of Factors in Allocating Funds and Selecting Projects

The Forest Service uses both quantitative and judgmental processes in deciding how to allocate fuel reduction funds. At headquarters, the agency increasingly relies on a quantitative process—reflected in a computer model—to determine the relative need for fuel reduction funds in each region. At the regional level, some offices primarily use quantitative processes to allocate resources while others rely on professional judgment. Similarly, the national forests use a mix of quantitative and judgmental processes to select projects.

At the National Level, the Forest Service's Allocation Process Increasingly Relies on a Quantitative Approach

At the headquarters level, the Forest Service has developed a computer model that assesses regions on various factors and assigns a score to each region reflecting its relative priority for fuel reduction funds.[22] According to Forest Service officials, they developed the model to address shortcomings that were highlighted by Congress and that were previously identified by GAO, the Department of Agriculture's Office of Inspector General, and the Office of Management and Budget. These shortcomings included inadequate assessment of the risk of wildland fires to communities, failure to clearly identify fuel reduction priorities, and little assurance that funding is targeted to these priorities.[23] In addition, agency officials said they developed the model to provide transparency, so that agency officials at all levels, as well as Congress and others, can understand the rationale behind allocation decisions.

In developing the model, the Forest Service brought together an interdisciplinary group of senior leaders to determine the final list of factors, which was based on an initial list developed by regional fuel program managers. To determine the factor weightings, the group followed a multistep process in which they determined the relative importance of each factor by comparing it separately to every other factor, and then synthesized the results to determine overall weightings.[24] The model includes 18 weighted factors, as shown in table 1.

Several of the factors—such as fire potential, ecosystem vulnerability, and wildland-urban interface—are designed to assess the potential for severe wildland fires in each region and the likelihood of damage resulting from such fires. For example, to determine the potential for severe fires, the model analyzes data such as the number and size of large wildland fires in each region. To determine the likelihood of damage resulting from wildland fires, the model includes data on values at risk, such as the locations of municipal water supplies and wildland-urban interface.

Other factors are intended to encourage efficiency and effectiveness within the fuel reduction program and across multiple Forest Service programs, such as the forest products or wildlife management program, to take advantage of opportunities to achieve objectives in other programs. Regarding effectiveness, the model included a factor intended to assess effectiveness in the regions—method availability. However, in practice, this factor used data on the total number of acres treated in each region— in effect rewarding regions for treating a large number of acres regardless of how well the treatments reduced risk or of the risk level of the areas treated. In addition, the model was designed to include a factor to assess how effective individual fuel reduction treatments are likely to be in reducing risk. However, the Forest Service does not currently have data to make such an assessment; consequently, for 2007, this factor did not influence allocations to the regions.

Table 1. Factors Considered in Forest Service Fuel Reduction Funding Allocation

Model Factors	Weight (percent)
Treatment effectiveness[a]	16.7
Wildfire potential	12.5
Wildland-urban interface	10.3
Treatment method availability	8.3
Wildlife habitat objectives	7.1
Municipal water supply	5.2
Ecosystem vulnerability[b]	5.2
Associated benefits[a,c]	4.2
Vegetative maintenance	4.2
Biomass opportunity	3.6
Insects and disease	3.6
Invasive species	3.6
Vegetation departure[b]	3.6
Watershed condition	3.6
Life cycle cost[a,d]	2.8
Commercial timber	2.6
Smoke emissions[e]	1.7
Use of legislative tools[f]	1.4
Total	**100**

Source: GAO analysis of Forest Service data.

Notes: Totals do not add to 100 percent due to rounding.

[a]No data were available for the 2007 allocation process.

[b]"Ecosystem vulnerability" and "vegetation departure"are measures of fire regime condition class.

[c]"Associated benefits" is a measure of acres treated with fuel reduction funds that achieve benefits for other programs, such as wildlife or watershed.

[d]"Life cycle cost" is intended to e the cost of treatments per year.

[e]"Smoke emissions" is a measure of the acres ofvegetation that produces high levels of smoke during a wildland fire.

[f]"Use of legislative tools" is ameasure of acres treated in projects authorized in HFRA or HFI, identified in community wildfire protection plans, or implemented using stewardship contracts. Stewardship contracting involves the use of any of several contracting authorities that were first authorized for use by the Forest Service on a pilot basis in 1998, and were subsequently extended to BLM. In practice, stewardship contracts generally involve the exchange of goods, such as timber, for contract services, such as thinning of brush.

In 2007, officials used the model's funding allocations to the regions, although they relied mainly on the prior year's funding levels the model's results only the model was still being refined and because they wanted to phase in funding changes gradually in order to minimize budget-related disruptions. Headquarters officials said they expect the model to have a stronger influence on future allocation decisions.

The model assigned a numerical score to each region that indicated the region's relative priority indicating higher priority. However, as shown in table 2 and figure 5, the Forest Service's funding the priority scores resulting from the model, with some high-scoring regions—such as the Northern and Eastern regions—receiving less funding than some lower scoring regions such as the Pacific Southwest and the Southwest regions.

Table 2. Forest Service Regions' Year 2007 Fuel Reduction Priority Scores and Funding Allocations

Dollars in thousands		
Forest Service region	**Priority score**	**Funding allocation**
Southern	574	$29,092
Northern	455	15,782
Eastern	416	9,718
Intermountain	408	16,165
Rocky Mountain	399	25,445
Pacific Northwest	389	25,794
Pacific Southwest	388	43,737
Southwestern	367	37,341
Alaska[a]	a	805
Total		**$203,879**

Source: Forest Service.

Notes: The Forest Service also allocated $2.265 million to its research stations and $ 95.109 million to its headquarters office and to cost pools, which are used for expenses that cannot reasonably be charged to a single program, including indirect, support, and common services charges.

[a]The Forest Service excluded the Alaska region from its model because the region has a small fuel reduction program relative to the other regions and receives less than 1 percent of the reduction funds.

According to Forest Service officials, the allocation amounts were not more closely correlated with the priority scores for the following reasons:

- As noted, the officials wanted to temper changes to regions' budget allocations until they completed revisions to the model and developed more confidence in its output in order to minimize funding shifts that might prove inappropriate once the model is refined.[25] Agency officials told us that even when they become confident in the model's output, they will likely implement changes incrementally in order to minimize disruption to regions and national forests.
- Until the revisions have been completed, the model's results will be tentative. An important focus of the revisions will be those 3 of the model's 18 2007. Because of the lack of data, these elements had no effect on the regions' [26] but agency officials hope to have data to2007 inform these elements for future allocations.
- The relatively high priority score assigned to the Eastern region was not consistent with agency officials' knowledge of the area-that is, they believed that, relative to the other regions, there were fewer destructive wildland fires in the Eastern region and, consequently, they expected the region's officials consulted data on the number of structures burned in wildland fires, their belief was confirmed. Consequently, agency officials are reexamining the measures they used to assess risk and exploring options for refining them.
- Fuel reduction costs vary widely from region to region, and when making final allocations, the officials made adjustments to accommodate this variation. For example, the Pacific Southwest region received the largest allocation of any region, despite its relatively low priority score, in part because treatment costs in the region are very high (averaging about $535 per acre in 2006). Therefore, a relatively large

allocation is needed to fund even a moderate amount of work. At the other end of the spectrum, treatment costs in the Southern region are low (averaging about $32 per acre in 2006), meaning that needed work can be accomplished with a smaller allocation. In addition, agency officials said that, although the Southern region's priority score might point toward a larger allocation for the region, nonmonetary constraints—such as the size of the workforce—limit the amount of work the region can accomplish and, therefore, the amount of funds that can prudently be invested there. Further, in order to maintain overall funding stability to the regions, officials coordinated regional fuel reduction funding allocations with those of other Forest Service resource programs, such as watershed management or forest products. This coordination sometimes resulted in officials adjusting fuel reduction funding allocations in order to compensate for adjustments in these other programs'

- The Forest Service allocated a portion of its fuel reduction funds according to congressional direction. Specifically, congressional committee reports accompanying relevant appropriations acts directed the Forest Service to spend about $34 million (12 percent) of its 2006 fuel reduction funds in certain areas or on certain projects. The Forest Service accommodated this congressional direction, regardless of whether doing so was consistent with priority scores.

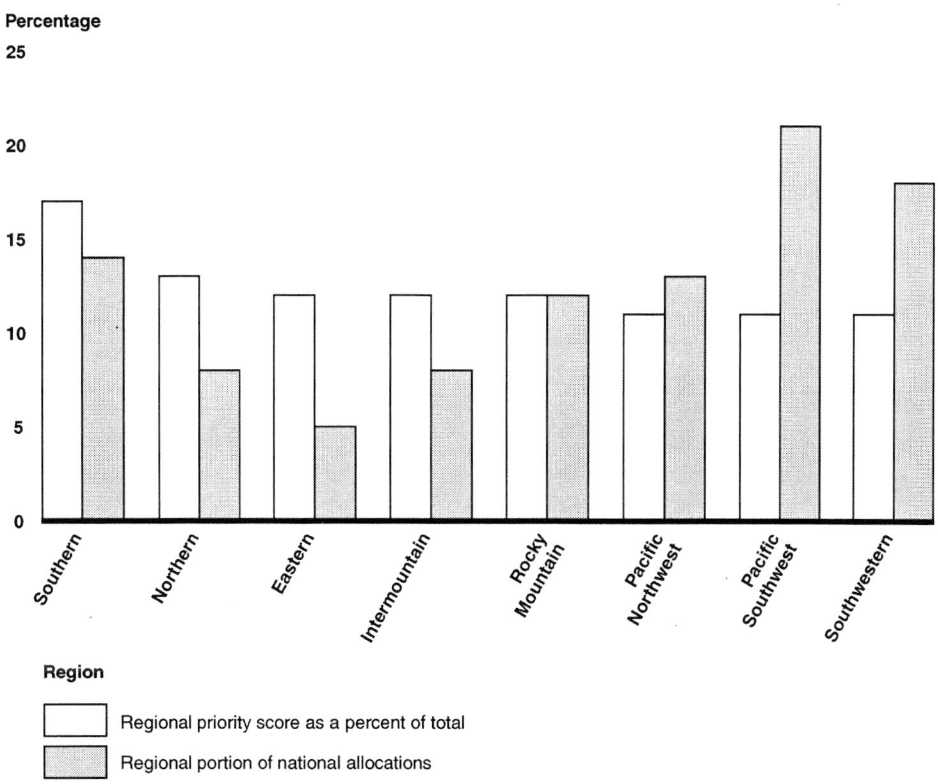

Source: GAO analysis of Forest Service data.

Figure 5. Forest Service Regions' Fueltion Priority Scores as a Percentage of Total, Compared to Regions' Funding locations as a Percentage of Total Allocations, Fiscal Year 2007.

Some Forest Service Regions Use Quantitative Allocation Processes, While Others Rely More on Professional Judgment

For 2007, the Forest Service allowed each region to determine how to allocate funds to its national forests and what factors to consider in the process, as long as the factors were consistent with those considered in the national allocation process. Four of the Service's nine regions relied primarily on quantitative data in their allocation processes to national forests, while five relied primarily on professional judgment. In applying these processes, all nine regions considered a combination of factors, many of which were similar to those used at the national level. Beginning with the 2008 allocations, the Forest Service plans to require regions to use the headquarters model to inform allocation decisions.

Of the four regions that relied primarily on quantitative processes in 2007, one—the Rocky Mountain region—used a computer model that analyzed geospatial data on vegetative condition and areas of insect-killed trees to help assess relative wildland fire risk among the national forests in the region. Regional officials then used their judgment to consider other factors, such as lands in the wildland-urban interface and acreage targets, to refine allocation amounts. Through the risk assessment process, the region identified 6 emphasis forests out of the 11 forests in the region and allocated over 70 percent of theregion's fuel reduction funds to these 6 forests. The Pacific Northwest region also used a model, but its model incorporated regional data on a number of factors, including the number of acres in fire regimes I, II, and III; the number of acres identified in community plans as being in the wildland-urban interface; and per-acre treatment costs. Using these data, regional officials identified five forests in the region where an extremely wet climate made the risk of damaging wildland fires so low that they decided not to allocate any fuel reduction funds to these forests and excluded them from the model. Another region that relied on a quantitative process—the Pacific Southwest region—used a scoring system that ranked forests primarily on the basis of a risk assessment; the assessment incorporated multiple factors, such as the number of acres in condition classes 2 and 3 and in the wildland-urban interface. The region also used other factors, such as the forests' capacity to conduct fuel treatment work, to make smaller adjustments. The Intermountain region allocated about 80 percent of its fuel reduction funds in accordance with forests' funds, the region delegated priority decisions to collaborative interagency groups in each of the region'ses. These groups scored and ranked proposed projects against a set of standard criteria and made funding recommendations to the regional office.

The remaining five Forest Service regions relied primarily on professional judgment and negotiation among agency officials when determining funding allocations to national forests. Although these regions did not use quantitative processes to assign priorities among forests, they incorporated some of the same information included in other regions' quantitative processes. For example, the Northern region conducted a risk assessment for the region, but instead of using the risk assessment to guide its allocations to the forests, the region directed forests to use it to identify potential treatments. The region then allocated funds to forests primarily on the basis of the forests' Southern region, officials used their professional judgment to decide on allocations largely on the basis of forests' reported capabilities, per-acre treatment costs, and local priorities, and how they fit with expected regional targets and budgets.

Factors outside of the formal process influenced allocations, according to Forest Service officials, but they did not always formally incorporate these factors into the allocation process. For example, in several regions, fuel reduction officials said they coordinated with officials from other resource programs, such as the wildlife management and vegetation management programs, when deciding on final allocations. In doing so, they sometimes adjusted fuel reduction allocations to, for example, prevent multiple programs from reducing allocations to a given forest in the same year or to take advantage of efficiencies when different programs' priorities overlapped in a given forest. In addition, nearly every region reported considering acreage targets when making allocation decisions—even those that did not report it as an official part of their allocation processes. Regional officials noted that pressure to meet the acreage targets established by Forest Service headquarters sometimes trumped all other factors in allocation decisions, especially in 2007 when targets increased at a faster rate than funding levels. Another factor, according to agency officials in several locations, was direction contained in congressional committee reports accompanying relevant appropriations acts that a certain amount of funding be allocated to specific forests or specific districts within forests. As with headquarters, regional offices allocated funds according to this direction, apart from any priority-setting process. For example, in 2006, the Pacific Southwest region allocated nearly $21 million (about 52 percent of the region's budget) on the basis of congressional committee report direction, in part to treat areas of insect-killed trees in southern California. Finally, regions reported shifting funds among forests, after initial allocation decisions had been made, to accommodate unexpected circumstances during the year, such as large wildland fires that prevented fuel reduction treatments from being implemented as planned.

National Forests Select Projects Using Quantitative and Judgmental Processes

Like regional offices, national forests are allowed to determine what processes to use and which factors to consider in selecting fuel reduction projects to fund and implement, within the parameters of national and regional direction. In practice, some forests rely more on quantitative, data-driven processes, while others rely more on professional judgment. Both consider a mix of factors, as the following examples, based on our site visits to national forests, illustrate:

- *Quantitatively based selection.* The Arapaho-Roosevelt National Forest in Colorado collaborated with another national forest, a national park, Forest Service research scientists, and the Colorado State Forest Service to develop a risk assessment that used quantitative data to map the highest priority locations for fuel reduction treatments in the area. Forest officials then used the risk assessment to prepare a 10-year strategy with proposed annual treatments. Each year, forest officials first consult the strategy and the risk assessment to identify a list of projects to fund, and then adjust the list to meet acreage targets within budget constraints. Similarly, officials of the Angeles National Forest in Southern California convened a diverse group of stakeholders and followed a step-by-step process to identify priorities for fuel reduction treatments. During the process, Forest Service officials provided information, such as the locations of historical wildland fires and developed areas, as well as places where fuel reduction was not feasible because, for example, the topography was too steep to operate needed equipment. They then used fire behavior models to show where fires could potentially burn and how various proposed fuel reduction treatments might affect such fires. The end result was a multiyear list of proposed projects that forest officials used to select projects each year.

- *Judgmental based selection.* At the Medicine Bow-Routt National Forest in Wyoming and the Chattahoochee-Oconee National Forest in Georgia, officials relied largely on their knowledge and experience about the area to select fuel reduction projects. Some of these officials had worked at the same forest for decades. At the Ocala National Forest in Florida, officials use their professional judgment to select projects, which are almost all prescribed burns. However, because the forest's fuel quickly, the project selection process is founded on a rotational schedule. Under this schedule, the forest aims to treat nearly all of its approximately 130,000 burnable acres over a 4-year period. Consequently, officials try to treat about a quarter of the acreage—or slightly over 30,000 acres—each year. Forest officials also said they consider other factors, such as wind direction, humidity, and human activity (for example, popular areas for weekend recreation), when determining the specific timing of a prescribed burn.

In addition to factors that national forests considered, unanticipated factors influenced project selection decisions at the local level, sometimes preventing planned projects from being implemented. In such cases, agency staff frequently carried out lower priority projects in place of the originally planned projects. For example, wildland fires sometimes burned in locations planned for fuel reduction treatments, making the treatments unnecessary; in other cases, litigation prevented planned treatments from being implemented as scheduled.

Interior and Its Agencies Use a Mix of Quantitative and Judgmental Processes and Consider a Range of Factors in Allocating Funds and Selecting Projects

Interior and its agencies—BLM, BIA, FWS, and NPS—use both quantitative and judgmental processes for allocating fuel reduction funds and selecting projects, and consider multiple factors, many of which are similar to those used by the Forest Service. In 2007, Interior allocated funds to its four agencies primarily on the basis of historical funding levels, but it is currently developing a computer model similar to the Forest Service's. Like Interior, the BLM national office allocated funds to its state offices primarily on the basis of historical funding levels in 2007 but is expecting to implement a new funding allocation model in 2008. The majority of BLM state and local offices allocated funds and selected projects using quantitative processes, many of which use scoring systems. The other three Interior agencies' quantitative and judgmental processes to allocate funds and select projects consider a range of factors.

Interior Allocates Funds to Its Agencies Primarily on the Basis of Historical Funding Levels

Interior's allocations to constant from year to year, measured on a percentage basis, because the department primarily allocates fuel reduction funds on the basis of past funding levels—what one departmental official called "allocation by tradition." This funding pattern dates back to 2001, when the Interior agencies began receiving a sharply increased amount of fuel reduction funds as a result of the National Fire Plan. Since then, the percentage of Interior's fuel reduction has remained consistent, with BLM receiving about 50 percent of the funding, BIA receiving about 20 percent, and FWS and NPS each receiving about 15 percent. Figure 6 shows the percentage of Interior's total fuel reduction appropriation that was distributed to each agency from 2001 through 2007.

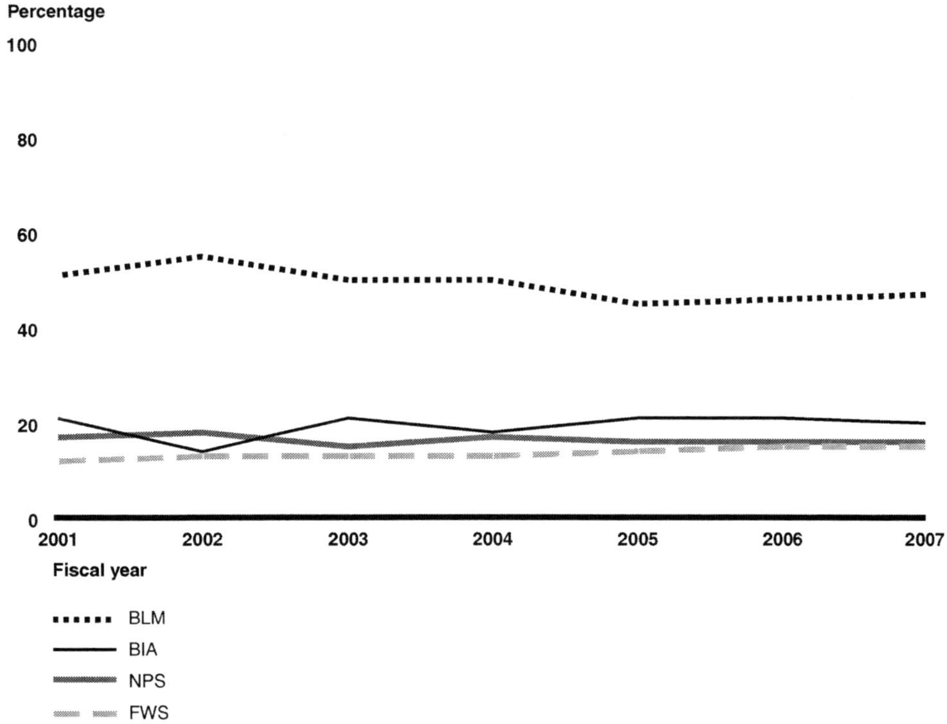

Source: GAO analysis of Interior data.

Figure 6. Percentage of Interior's Total Fuel Reduction Funds Allocated to the Interior Agencies, Fiscal Years 2001 through 2007.

In 2001, Interior established initial funding allocations on the basis of estimates of each agency's infrastrucure and capacity (i.e., the amount of work each could accomplish), which it determined by compiling field requests from the four agencies. However, at the time, most of the agencies and their field units had little infrastructure related to the fuel reduction program—including limited staff—so many units did not have the resources to collect extensive information on fuel reduction needs, according to agency officials. As a result, agency officials had to make allocation decisions based on limited information. Some agency and departmental officials have stated that the allocations need to be revisited now that the fuel reduction program has been in place for several years.

Each year, the department tells the four agencies how much funding the department has requested for the fuel reduction program and what its acreage targets are for treatments within and outside of the wildland-urban interface. The agencies' officials in charge of each agency's determine how to divide the funds and set targets for each agency. However, the fuel program leads do not have the authority to significantly adjust the funding allocations from previous levels; rather, such changes would have to be determined at the department level, according to headquarters officials. The agencies' lists to the regions, which review these lists before forwarding them to headquarters; these lists provide the fuel program leads with an idea of each agency's needs allocations. After the fuel program leads decide upon initial allocations, they may shuffle funds within or between their agencies throughout the year to adapt to uncontrollable circumstances, such as weather conditions. The majority of fuel treatments that the Interior agencies conduct depend on the weather, and sometimes weather conditions prevent work from being completed. For example, if a drought in the Southeast makes vegetation too dry for safe prescribed burns, Interior may shift funds to units in the western United States. In practice, these considerations may result in Interior's which conduct a large number of fuel treatments in the Southeast, to BLM or BIA, which conduct most of their fuel treatments in the West.

In 2007, Interior allocated 5 percent of its funds to the agencies using a model similar to the Forest Service model, and it plans to use the model to influence a greater portion of allocations in future years.[27] The department developed the model after Interior and the Forest Service received congressional committee direction in 2005 to develop a common method for setting project priorities. Interior's 2007 model included a range of factors, such as the amount of land each agency manages with certain fuel conditions and the degree to which each agency used biomass, but included fewer factors than the Forest Service's model because some data were not yet available. Because Interior does not currently have a good method for measuring efficiency or effectiveness, its 2007 model used the legislative tools factor, which measures the extent of use of HFI and HFRA planning authorities, to measure efficiency, and the number of acres treated to measure effectiveness. The following provides the complete list of factors used in Interior's 2007 model. [28]

- number of fire starts,
- number of large fires (defined as 500 acres or more),
- fuel conditions,
- biomass utilization,
- number of threatened and endangered species,
- fire regime condition class improvement,
- use of legislative tools (HFI/HFRA),
- number of acres treated, and
- wildland-urban interface.

Results from the 2007 model were generally consistent with Interior's allocations to the agencies in previous years. However, Interior is still making changes to the model, including determining how to weight the factors, so this may not be the case in future years. According to departmental officials, Interior intends to be cautious in applying the new model and making significant changes to current allocations because Interior and the Forest Service are currently developing the Fire Program Analysis (FPA) system—an interagency fire management planning and budgeting model—and they expect information from that system to inform future allocation decisions.[29] By proceeding slowly, the department hopes to avoid potentially disruptive fluctuations in regional and field unit allocations.

Once they have received their allocations from the department, Interior agencies determine how to allocate fuel reduction funds to the regions within the parameters of departmental and congressional direction. Interior officials have stated that they would like the agencies to use more rigorous allocation processes in the future, though one departmental official noted that he does not want the agencies to invest substantial funding or time and effort to develop new allocation processes pending the expected completion of the FPA. Interior guidance lists the following priorities for selecting projects:

- All projects must result from a collaborative process.
- Funding will be targeted to the wildland-urban interface.
- Within the wildland-urban interface, focus should be on projects near wildland-urban interface communities at greatest risk of fire; communities that have completed a community plan or its equivalent; and communities where there is an active partnership with volunteer efforts, in-kind services, or partners who contribute funding.
- Outside of the wildland-urban interface, focus should be on areas in condition class 2 or 3 in fire regimes I, II, or III, or those in condition class 1 where landscape conditions could quickly deteriorate to condition 2 or 3.
- Priority should also be given to projects using mechanical treatments, with special emphasis on projects yielding biomass that can be sold or traded to companies or the local community; and projects using contractors, particularly those projects conducted under contracts that support rural communities'
- Prescribed burning is to be used when weather and resource conditions permit, where mechanical treatments are not appropriate, and as maintenance treatments following mechanical work.
- Managers must make maximum practical use of tools provided by HFRA and HFI.

BLM Increasingly Uses Quantitative Processes in Allocating Funds and Selecting Projects

In 2007, BLM headquarters allocated funds to its state offices primarily on the basis of historical funding levels; however, agency officials told us that, starting in 2008, BLM plans to use a quantitative process incorporating factors similar to those used in Interior's new model, with a greater emphasis on collaboration and local priorities. BLM headquarters provides flexibility to state offices and local units when allocating funds and selecting projects but directs these offices to consider Interior and agency guidance. The majority of BLM state offices and local units used quantitative processes to allocate funds and select projects in 2007, frequently scoring projects against a set of weighted factors.

BLM Allocates Funds to Its State Offices Primarily on the Basis of Historical Funding Levels but Plans to Use a More Quantitative Approach in 2008

In 2007, BLM headquarters allocated funds to its state offices largely on the basis of past funding levels—as in previous years—as a way to ensure that funding levels remain relatively stable, but it also considered proposed projects, national priorities, and the extent to which state offices met past acreage targets established by BLM. While the project lists do not largely influence allocations to state offices, state offices use these lists to allocate funds to field units, and field units use them to select projects for implementation. Table 3 shows the 2007 allocations to the BLM state offices. (App. II also shows 2005 and 2006 allocations.)

Table 3. BLM Allocations to State Offices, Fiscal Year 2007

State office	Allocation	Percent of BLM total state office allocation
Oregon/Washington	$24,878,000	27.1
Idaho	14,598,000	15.9
Utah	10,078,000	11.0
California	7,322,000	8.0
Colorado	6,843,000	7.5
Nevada	6,414,000	7.0
New Mexico	6,412,000	7.0
Montana	5,461,000	6.0
Arizona	4,355,000	4.7
Wyoming	3,684,000	4.0
Alaska	1,556,000	1.7
Eastern States	126,000	0.1
Total	**91,727,000[a]**	**100.0**

Source: GAO analysis of BLM data.

Notes: Total allocation includes the allocation for the current year plus carryover from the previous fiscal year.

[a]BLM allocated an additional $8,473,000 for BLM headquarters, science centers, training costs, and other support costs.

As shown in table 3, the Oregon/Washington, Idaho, and Utah state offices got substantially more funding than the other states—more than half of BLM's total funding. more than $24 million—27 percent of BLM's state office funding, one BLM field unit in Oregon, the Medford district office, received over $9 million in 2007—more than nine state offices each received in total funding. According to some agency officials, the relatively high level of fuel reduction funding directed toward the Oregon/Washington state and Medford district offices is, in part, the result of BLM's emphasis on providing stable levels of funding to states and field units. According to these officials, when BLM (along with other federal agencies) received a sharp increase in fuel reduction funding in 2001, agency officials sought to identify units that could implement fuel reduction projects quickly. Because the Oregon/Washington and Medford offices were identified as having the capacity to undertake a large number of fuel reduction projects, they received a substantial portion of the new funding. However, another agency official told us these large amounts are justified because there is substantial wildland fire risk in Oregon and, therefore, a great need for fuel treatments because vegetation grows very quickly in the western part of the state, there is considerable wildland-urban interface, and wildland fire suppression costs are high.

Starting in 2008, BLM plans to use a model to influence funding allocations to state offices for fuel reduction. Use of the model is intended to ensure that the highest priority work is funded and that BLM's fuel reduction treatments are integrated with other vegetation treatments, such as range improvement projects, to effectively achieve fire and resource management goals and objectives. According to a headquarters official, the new model is intended to facilitate comparison of risk and needed work at the national and state levels in order to set priorities for funding among states and communities. Headquarters officials will use the model results to make allocation decisions but will shift no more than 20 percent of the previous year's allocations to each state in 2008 and 2009.

The model has three components: (1) treatment characteristics; (2) a measure of the degree of threat; and (3) an efficiency measure. For the first component—treatment characte ristics—the model will score every proposed project on a set of weighted factors, such as local priority ratings, the availability of joint funding, and condition class; there are separate factors and weights for projects within and outside of the wildland-urban interface. The second component—the measure of the degree of threat—currently combines three elements: the number of fire starts, the number of large fires (i.e., fires greater than 300 acres), and local risk ratings. The third component—efficiency—is currently measured by past performance on acreage targets, past performance on estimating treatment costs, and treatment cost per acre. According to agency officials, they intend to eventually include a measure of effectiveness in the model, which would indicate how well a treatment reduces risk or achieves other objectives. However, because BLM does not currently have a good way to measure effectiveness, it is using measures of efficiency until it develops a better approach. Table 4 shows the complete list of factors used in the model and their weights.

Table 4. Factors and Factor Categories BLM Considers in BLM Fuel Reduction Funding Allocation Model

Funding allocation model components	Factors evaluated	Weights for wildland-urban interface treatments	Weights for treatments outside the wildland-urban interface	Overall weight
	Community plan or equivalent	0.14	0.01	
	High local priority[a]	0.14	0.11	
	Mechanical treatment	0.12	0.04	
	Joint funding available	0.10	0.08	
	HFRA/HFI NEPA type[b]	0.10	0.08	
	Stewardship project[c]	0.10	0.08	
	Multiple land ownership	0.08	0.03	
Treatment characteristics	Moderate local priority[a]	0.08	0.06	**0.45**
	Biomass utilized[d]	0.05	0.04	
	Large-scale treatment[e]	0.05	0.10	
	Low local priority[a]	0.02	0.02	
	Condition class 2 or 3	0.01	0.11	
	Impacted species	0.01	0.10	
	Fire regime I, II, or III	f	0.06	
	Fire or other treatment method[g]	f	0.08	

Table 4. (Continued)

Funding allocation model components	Factors evaluated	Weights for wildland-urban interface treatments	Weights for treatments outside the wildland-urban interface	Overall weight
Degree of threat	Number of large fires (greater than 300 acres)[h]	0.50	0.50	0.35
	Number of fire starts[h]	0.25	0.25	
	Local risk rating[i]	0.25	0.25	
Efficiency	Past performance on acreage targets[j]	0.50	0.50	0.20
	Past performance on treatment cost estimates[k]	0.40	0.40	
	Cost per acre	0.10	0.10	

Source: GAO analysis of BLM data.

[a]The local priority rating is assessed at the local level and is a way for the field to communicate project priorities that may not be well-represented by other factors.

[b]The "HFRA/HFI NEPA-type" factor weights projects that use National Environmental Policy Act (NEPA) planning tools authorized by HFRA or HFI.

[c]Stewardship projects are accomplished through the use of stewardship contracting, which involves the use of any of several contracting authorities that were first authorized for use by the Forest Service on a pilot basis in 1998, and were subsequently extended to BLM. In practice, stewardship contracts generally involve the exchange of goods, such as timber, for contract services, such as thinning of brush.

[d]The "biomass utilized" factor weights projects hat make use of biomass—small-diameter trees, branches, and other organic material—removed through fuel reduction.

[e]Large-scale treatments are treatments that are at least 150 percent larger than the average treatment.

[f]This factor was not used to determine the treatment scores for wildland-urban interface treatments.

[g]This factor weights projects treated with prescribed fire or other treatment methods, such as grazing or herbicides.

[h]The "number of large fires" and "number of fire starts" factors are determined at the field office or district level and applied to all treatments within that field office or district.

[i]The risk rating is assessed at the local level and is to be determined from community plans or risk assessment programs.

[j]The "past performance on acreage targets" factor is calculated at the state level and applied to all treatments within the state.

[k]The "past performance on treatment cost estimates"factor is calculated at the state level and applied to all treatments within the state.

The BLM national office also directs state offices and local units to consider Interior and BLM priorities when allocating funds and selecting projects. BLM-specific guidance directs state offices and local units to coordinate fuel treatments with other resource management activities, such as timber and wildlife habitat; target funds to wildland-urban interface areas identified through a collaborative process; target non-wildland-urban interface funds to ecosystems that have the highest risk-reduction potential; and use HFI and HFRA planning tools.

The Majority of BLM State Offices Incorporate Quantitative Approaches in Their Allocation Processes

The BLM national office allows state offices to choose the approach they use in allocating funding to field units, as long as they take into account departmental and BLM priorities, and state offices will continue to have this flexibility with the implementation of the new national allocation process, according to headquarters officials. In 2007, 6 of the 11 BLM state offices primarily used quantitative approaches to inform their allocation processes, and 5 primarily used a judgmental approach.[30] Nine of 11 state offices considered targets or past performance, and 10 considered at least one factor related to collaboration, such as community plans. Eight of 11 state offices considered at least one factor to estimate wildland fire risk, such as local- or state-level risk assessments or fire regime condition class.

The six state offices that allocated funds using quantitative processes in 2007 primarily used weighted scoring systems—similar to the state scoring component of BLM's specific factors and their weights varied by state, many factors were commonly used and were similar to those used in BLM's headquarters system; each of the states had separate lists of factors for projects within and outside of the wildland-urban interface. For wildland-urban interface projects, five of the six state offices emphasized factors such as local risk ratings or community hazard assessments to estimate risk from wildland fire, and all six offices considered a variety of other factors, including community support and joint funding, to measure the extent of collaboration. For projects outside of the wildland-urban interface, all six offices gave priority to projects in condition classes 2 or 3, jointly funded or collaborative projects, and projects that improved threatened and endangered species habitat, as well as a variety of other factors. Once the state offices had the field offices project lists, state and field offices generally negotiated to determine final funding allocations.

The remaining five state offices primarily used judgmental processes to allocate funding to field units. For example, in 2007, the Oregon/ Washington state office allocated funding using professional judgment and negotiation, which included numerous discussions with field units' fuel program staff to assess the units' priorities and capabilities. The state office primarily considered capability and past performance of field offices and BLM's national 2008, the office plans to use a model to allocate base funding for fuel reduction, which covers salaries and other fixed costs, but will continue to allocate project funding using the current approach, which relies primarily on professional judgment and negotiation.

BLM officials told us that factors outside of the formal process influenced allocations. For example, in several states, agency officials said they coordinated with officials from other resource programs, such as the range or weeds programs, at the state or local level when deciding on final allocations or selecting projects. As a result, they sometimes selected projects that used funding from multiple resource areas, or benefited these areas, over other projects in order to take advantage of efficiencies. Many state offices also reported that they considered acreage targets when making allocation decisions. According to one state official, acreage targets were the most influential factor in allocation decisions, and several agency officials said that lower priority projects were sometimes funded to meet acreage targets. Finally, state offices reported shuffling funds among or within field units after allocation decisions had been made to accommodate uncontrollable circumstances throughout the year, such as weather conditions that prevented prescribed burns from being implemented as planned.

BLM state offices also devote substantial effort and funding to assist in the development of community plans, and allocate a significant portion of fuel reduction funding to projects on private land. For example, the Montana state office funded the development of 49 out of 54 completed community plans throughout the state, according to an agency official. Also, when the Montana state office and its field units allocate funding to, and select projects in, the wildland-urban interface, proposed projects on private land—which are submitted to BLM by counties—are ranked using the same system as BLM projects. Consequently, BLM projects on federal land essentially compete for the same funding as projects on private land. In California, the BLM state office allocates more than half of its wildland-urban interface funding to a community assistance program, through which fuel treatments on private, state, or tribal lands adjacent to or in the vicinity of federal lands are funded through an interagency grant process.

The Majority of BLM Field Units Incorporate Quantitative Approaches into Their Project Selection Processes

As with the BLM state offices, in 2007, the majority of BLM field units used quantitative approaches that incorporated a range of factors—many of which were similar or identical to the ones used by state offices—to select and rank projects. For example, the Twin Falls district office in Idaho scored all projects using a weighted scoring system developed by the BLM Idaho state office. It then ranked the projects, considering factors such as project scores and areas identified in community plans. The Billings field office in Montana also used a weighted scoring system to rank projects. Field staff initially identified projects using community plans or the field office's risk assessment class, and fire occurrence to identify high-risk fire areas—and then scored the projects using the Montana state office's weighted scoring system to identify high-, medium-, and low-priority projects.

Also, like national forests, BLM field units were sometimes influenced by unanticipated factors when selecting projects. For example, agency officials sometimes deferred planned projects because newly proposed projects suddenly became a high priority. They pointed to situations in which nonprofit organizations donated funds to pay for projects and agency officials gave those projects a higher priority. In Colorado, recent oil and gas development, as well as construction of new subdivisions in high-risk areas, have caused field units to shift priorities to conduct treatments near these developments, according to a BLM official.

BIA Allocates Funds Largely on the Basis of Unit's Performance History, while FWS and NPS Use Quantitative and Judgmental Processes

In 2007, the three remaining Interior agencies—BIA, FWS, and NPS— allocated fuel reduction funds using quantitative and judgmental processes and considering a variety of factors. Like BLM, these agencies provide performance flexibility to regional offices and local units in determining how to allocate funds and select projects and direct them to consider departmental priorities. (See app. II for these agencies' 2005 through 2007 allocations to their regional offices.)

BIA and FWS Headquarters Allocate Funds Using Quantitative Processes, While NPS Headquarters Allocates Funds Primarily on the Basis of Historical Funding Levels

In 2007, BIA headquarters allocated fuel reduction funds to its regions using a formula that considered past performance and proposed work and that essentially rewarded regions for their accomplishments. The formula allocated to each region a percentage of the region's total budget request, based on the percentage of the prior 3 years' acreage targets that the region met. For example, if a region had met 95 percent of its total acreage target since 2004, the region would receive about 95 percent of its requested budget for 2007. BIA placed a cap on the amount of funding that regions could request, based on their previous year's accomplishments.[31] According to a headquarters official, BIA rewards those regions and units that achieve acreage targets because, in many instances, units do not meet targets.

FWS headquarters allocated 2007 funds to regional offices using a quantitative model that considers multiple factors, including historical fire occurrence, fuel conditions, community assessments of risk, and field unit past performance. The model has separate modules for projects within and outside of the wildland-urban interface, and produces a weighted score for each FWS field unit. In the wildland-urban interface module, the most influential factors are communities at risk, local hazard rankings, and fire conditions. For the non-wildland-urban interface module, the most influential factors are past performance and proposed work.

NPS headquarters allocated 2007 funding to regional offices primarily on the basis of historical funding levels. These levels were originally set by a model that determined funding allocations through a risk assessment, which considered vegetation types, fuel types, fire return intervals, and other data, and through an effectiveness measure that examined treatment success for different vegetation types. According to an NPS official, the agency maintained funding proportions at the model's 2005 level after

Interior directed it to work on the FPA; NPS decided that it would have been too much work for field staff to maintain the model while also preparing data for the FPA. Furthermore, they believed that the model, initially developed more than 20 years ago, was outdated and did not merit additional financial investment while the FPA was being developed.

BIA, FWS, and NPS Regional Offices Allocate Funds to Field Units Using Quantitative and Judgmental Processes

BIA, FWS, and NPS allow their regions the flexibility to determine how to allocate funding to field units, provided the processes and factors are consistent with departmental and agency guidance. The BIA national office encourages regions to adopt allocation strategies similar to the one used at headquarters—which reward s past performance—and some of BIA's regional such as the Rocky Mountain and Northwest regions. The Rocky Mountain region, for example, used a quantitative process to allocate fuel reduction funds, using an allocation formula similar to the one used by BIA headquarters but using only the previous year's headquarters used. Likewise, one FWS regional office that we visited used a quantitative process to allocate fuel reduction funds in 2007: FSW's Mountain-Prairie region allocated funds to local units using FSW's national model, but regional officials adjusted the model's allocations on the basis of their knowledge about local factors, such as community support for projects and field unit staffing levels.

Other BIA and FWS regional offices, and all of the NPS regional offices that we visited, allocated fuel reduction funds in 2007 using judgmental processes that incorporated a range of factors. For example, BIA's Southwest region allocated funds primarily on the basis of project rankings (as determined at the local level) and cost efficiency, according to a regional official. Likewise, FWS's Southeast region allocated 2007 funds according to a regional such as field units' has many years of experience managing the region's fuel reduction program. In NPS's and fuel program staff determined funding allocations on the basis of park priorities, past performance, and conformance with NPS policy, balanced against regional funding levels and acreage targets.

As in other agencies, officials told us that factors outside of the formal process also influenced allocations. Several BIA and NPS officials told us that staffing constraints at field units may affect allocations. For example, many park units have very small fuel treatment programs and no staff dedicated solely to the program; therefore, the fuel reduction programs at such units may be eliminated if staff, who have numerous collateral duties, no longer have the time to plan or implement treatments. Furthermore, the location of some field units makes it difficult to recruit and retain qualified staff; the field units are located either in areas with high costs of living or in remote areas. Without dedicated staff to manage fuel reduction programs at such field units, their capacity to plan and implement projects and spend any funding allocation is limited, so capacity becomes the determining factor regardless of other factors considered in the allocation process, according to agency officials. Some BIA officials told us that self-determination limits BIA'sover the tribes; self-determination provides tribes with the authority to manage federal programs when they choose to do so, as well as the authority to choose not to emphasize a given program. In some regions, acreage targets also affected allocation and project selection processes, and one agency official told us that projects were sometimes developed and implemented specifically to meet targets. However, other BIA, FWS, and NPS regional officials told us that they did not assign acreage targets to field units or that there was little pressure to meet targets. Finally, regions reported shifting funds among field units after allocation decisions had been made to adapt to uncontrollable circumstances, such as weather conditions that prevented planned projects from being implemented.

Local BIA, FWS, and NPS Units Select Projects Using Quantitative and Judgmental Processes

Some BIA, FWS, and NPS local units selected projects in 2007 using quantitative processes. For example, in NPS's Sequoia and Kings Canyon National Parks in California, agency officials identified projects using a model that determined high-risk areas on the basis of several factors, such as the risk of a fire starting and the location of the wildland-urban interface. Park officials used the model information, as well as additional factors, such as values at risk, sequencing of treatments, and project accessibility, to select projects. used a quantitative process to select projects. BIA's Zuni Agency in New Mexico also used a quantitative process to select projects. The fuels specialist analyzed geographic information—for example, on housing density and existing vegetation—to identify and rank projects.

Other BIA, FWS, and NPS units primarily used judgmental processes when selecting projects for 2007. For example, at FWS's Merritt Island National Wildlife Refuge in central Florida, field staff selected projects primarily on the basis of the rotational schedule for prescribed burns. Refuge officials also considered other factors, such as wildlife habitat, to select which projects to complete that year. According to agency officials, the refuge has habitat for the scrub jay, a threatened species, and while prescribed burns generally improve this habitat, too much prescribed burning can be disruptive. NPS's Cape Canaveral National Seashore, which neighbors Merritt Island National Wildlife Refuge, also selected projects judgmentally, and in coordination with refuge staff. The process was primarily influenced by the location of the wildland-urban interface and threatened and endangered species habitat.

BIA, FWS, and NPS field units, like national forests and BLM field offices, also adapted to unanticipated events when selecting projects. In some cases, field units were forced to accommodate unique circumstances. For example, the Merritt Island National Wildlife Refuge is adjacent to a National Aeronautics and Space Administration facility, and, during the days immediately before and during scheduled rocket or shuttle launches, the refuge must put all prescribed burns on hold.

Several Improvements Could Help Better Ensure That Fuel Reduction Funds Are Allocated to Effectively Reduce Risk

Although the Forest Service and Interior are taking steps to enhance their funding allocation and project selection processes—for example, by developing models to assist in making allocation decisions—there are several improvements they could make to better ensure that they allocate fuel reduction funds to effectively reduce risk. Specifically, when allocating funds and selecting projects, the agencies could improve their processes by (1) consistently assessing all elements of wildland fire risk, including hazard, risk, and values; (2) developing and using measures of the effectiveness of fuel reduction treatments; (3) using this information on effectiveness, once developed, to assess the cost-effectiveness of potential treatments; (4) clarifying the relative importance of the numerous factors they use in allocating funds, including factors unrelated to risk or effectiveness; and (5) following a more systematic process in allocating funds. While the agencies have recognized the importance of these elements—particularly risk, treatment effectiveness, and cost effectiveness—in several strategy documents, they have not effectively incorporated them into their allocation processes.

The Agencies Do Not Consistently Assess All Elements of Risk When Allocating Funds

The agencies have repeatedly stressed the importance of identifying high-risk areas in setting priorities and allocating funds for fuel reduction; for example, in their 2006 document Protecting People and Natural Resources: A Cohesive Fuels Treatment Strategy (Cohesive Strategy),[32] the Forest Service and Interior declared that they "expect to ensure that fuel project investments are cost-effectively allocated to achieve risk reductions." Similarly, in its 2007 budget justification, the Forest Service declared that the fuel reduction program focuses on reducing the risk of wildland fire and long-term damage to resources and property; likewise, Interior's eclared that the department intended2007 budget to reduce fuels in order to "provide better risk reduction to communities and resources."

At the national level, the Forest Service and FWS headquarters incorporated nationwide risk assessments into their 2007 allocation processes; Interior did so for only 5 percent of the funds it allocated to the four Interior agencies; and BIA, BLM, and NPS did not include risk assessments in their national allocation processes at all, although BLM officials said they are taking steps to do so in the future. According to Forest Service and Interior agency officials, it has been difficult to develop national risk assessments because they require nationally consistent data, which have not always been available.[33] Furthermore, some of the available national data on vegetation type and condition were designed for forests and, consequently, are not as accurate for shrublands and grasslands.

At the regional and local levels, some agency offices used risk assessments when allocating funds and selecting projects, while others did not. One of the Forest Service's 9 all three elements required for a risk assessment in their 2007 allocation processes, and several other Forest Service regions considered two of the three elements—hazard and values—but did not consider risk. Some, but not all, of the other Interior agencies' considered elements of risk assessments in their allocation processes as well. Regional officials offered several reasons for not always systematically considering risk assessments when allocating funds, such as not having the necessary data for a regionwide risk assessment or only informally considering risk.

Several agency officials told us that they do not consider the lack of a formal national or regional risk assessment to be a significant problem because they rely on field units to assess risk when selecting projects. However, as with regions, not all local units used risk assessments when selecting projects; some local units used only partial assessments or did not use risk assessments at all. Even when field units do use risk assessments to help select projects in high-risk areas at the local level, agency officials cannot be confident that areas designated as high risk locally would still be designated as high risk at the regional or national level. For example, one BLM field office in Colorado oversees a rural area with only two communities, neither of which is at risk from wildland fire, according to BLM officials. For officials at this office, the most important values at risk are rural power lines and oil and gas infrastructure; therefore, they give the highest priority to projects that protect these features. From a regional or national perspective, however, other projects may be a higher priority for funding because the values at risk are more important, the area is at higher risk from fire, the level of hazard is greater because of fuel conditions, or some combination of these reasons. Without using national, regional, and local-level risk assessments that systematically assess hazards, risks, and values, it is difficult to ensure that allocation decisions are grounded in a clear understanding of which areas are at the highest risk.

Even when the agencies conduct risk assessments that include hazards, risks, and values, they may find it difficult to distinguish between high- and low-priority locations because one key value at risk—the wildland-urban interface—has multiple definitions that leave considerable room for interpretation on the part of agency officials. As a result, many different areas can be classified as wildland-urban interface, and the term's usefulness in helping agency officials identify, and direct funds toward, the highest-priority lands is diminished. In 2001, the agencies—in cooperation with tribes and states—defined the interface as including three categories: (1) dense populations (250 or more people per square mile) abutting wildlands; (2) scattered populations (28 to 250 people per square mile) intermixed with wildlands, and (3) development surrounding an island of wildland fuel, such as a park or open space. Agency officials told us that they developed this definition very quickly, in response to legislative direction, but later came to believe that it overemphasized population density and was not flexible enough to accommodate differences in landscape features such as vegetation, terrain, and prevailing weather patterns, which can affect the size and shape of areas in the wildland-urban interface.

In 2003, HFRA defined the wildland-urban interface to include an area within or adjacent to an at-risk community, that is identified in project recommendations to a federal agency in a community wildfire protection plan. For areas not in community plans, HFRA specified that areas within one-half mile of an at-risk community were to be considered wildland-urban interface,[34] as were areas within 1-1/2-miles of an at-risk community under certain conditions, and areas adjacent to evacuation routes for at-risk communities. According to agency officials, this definition offered more flexibility by moving away from the focus on population density, but it applies only to projects conducted using HFRA authorities.

Most recently, the 2006 10-Year Strategy Implementation Plan developed by the agencies, western governors, and others, defined the wildland-urban interface as the "the zone where structures and other human development meet at-risk forest and rangelands." While this definition provided broad flexibility, agency officials told us it did not replace the 2001 definition (which focused on population densities), and both the 2001 and 2006 definitions apply to projects other than those conducted using HFRA authorities. The end result is multiple definitions that—individually and collectively—allow many different areas to be classified as wildland-urban interface without specifying whether some ought to be given higher priority than others.

In part because of this lack of clarity, agency officials we spoke with reported including several types of locations under the category of wildland-urban interface. Some units interpreted the interface to mean only the area surrounding houses, while others also included roads, power lines, oil and gas development, communications infrastructure, campgrounds and recreation areas, and other features. For example, the BLM Colorado state office defined industrial interface as a subcategory of the wildland-urban interface, including features such as power lines or oil and gas development, which are common features in or near some of BLM's rural field urban areas (Atlanta and Los Angeles) determined that most or all of their forests were in the wildland-urban interface because, they estimated, a wildland fire could move into nearby urban and suburban areas within a single day. In yet another interpretation, a BIA agency in New Mexico tailored its definition of wildland-urban interface to accommodate cultural differences between tribes, as the differences were reflected in the arrangement of their homes: one tribe built its homes in clusters while another built its homes in a scattered pattern.

Although each of these interpretations of wildland-urban interface may have merit given the situations the field units face, the lack of clear definition effectively allows a wide range of areas to be defined as wildland-urban interface. The fluid nature of the wildland-urban interface definition is illustrated by guidance that one FWS region issued to its local units in 2006, when it notified them that it was expanding the relatively strict definition of wildland-urban interface the region had previously used to reflect interagency guidance. According to this region, "this expanded definition may enhance our ability to fund a project with [wildland-urban interface] ... funding and will help us meet the [wildland-urban interface] treatment targets mandated by the Department."

Given the range of definitions available for wildland-urban interface, it is not surprising to find that in 2005 and 2006 many of the fuel reduction treatments the agencies identified as being in the wildland-urban interface were in ZIP code areas with fewer than 28 people per square mile, on average.[35] (See figure 7.) Specifically, about 2.2 million acres, or 65 percent of all acres treated in areas identified as the wildland-urban interface during that period, were in ZIP code areas with fewer than 28 people per square mile. While the agencies may have had legitimate reasons for some of these treatments—for example, to protect a critical evacuation route for a larger community—it is not clear why, as a whole, so many acres treated are far from more densely populated areas. Expressing its concern about this situation in 2006, the Office of Management and Budget noted, "As the agencies increase their emphasis on [wildland-urban interface] treatments over time, field staff and/or project proponents may simply be defining more projects as [wildland-urban interface] projects in order to increase the likelihood of having their projects funded."

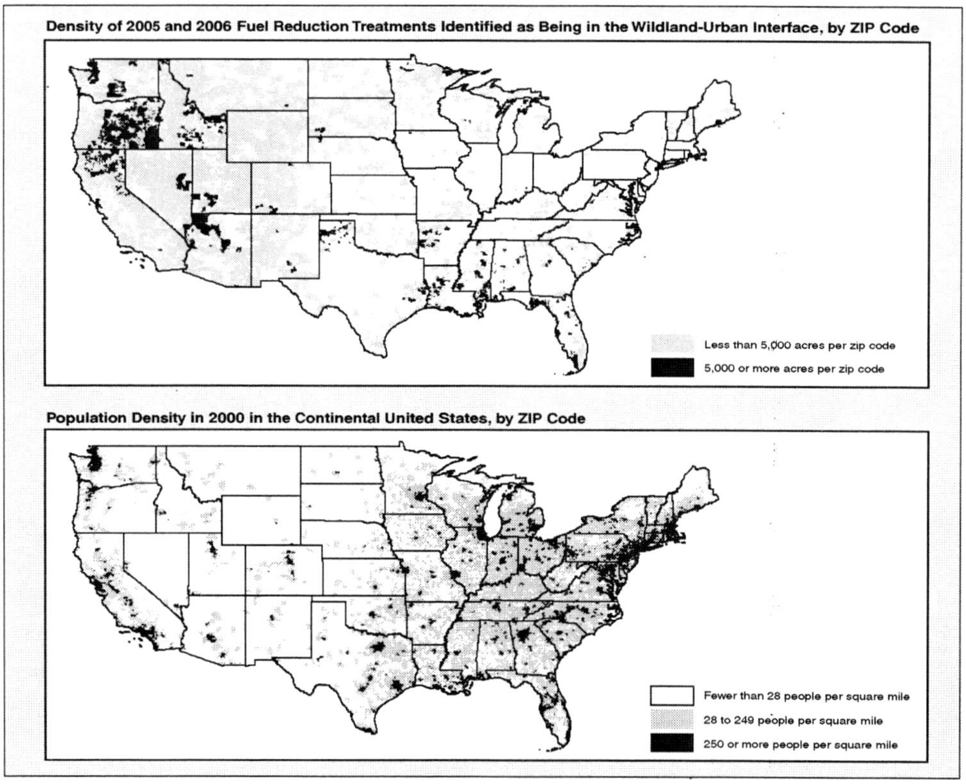

Source: GAO analysis of Forest Service, Interior, and U.S. Census data.

Note: We conducted our analysis using census data on the average population per square mile across areas defined by ZIP codes. However, especially in larger ZIP codes, there may be smaller pockets where the population density is higher or lower than the average used in our analysis. When mapping the data, we included a 1.5-mile buffer around the ZIP code areas with 28 or more people per square mile to account for the 1.5-mile buffer specified in the HFRA definition for wildland-urban interface.

Figure 7. Density of Wildland-Urban Interface Treatments and Population Density, by ZIP Code.

Conversely, population density alone may not be sufficient justification for selecting locations for fuel reduction. One Forest Service official cautioned against "prioritization by census," because more densely populated areas are not necessarily at greater risk than less populated areas. For example, although Chicago is a densely populated urban area, the Forest Service has not conducted more treatments in the nearby grassland because the risk of a fire threatening the urban area is very low, according to agency officials. In addition, highly populated urban areas are often not as close to federal lands as are communities with smaller populations, and the agencies conduct the majority of their fuel reduction work on federal lands. Figure 8 shows the location of federal lands relative to more densely populated areas in the continental United States. Even if a dense urban area is near federal lands, the entire area is not typically at risk from a fire originating on federal lands; only the portion of structures closest to federal lands is at risk, according to Forest Service officials.[36] Finally, vegetation and other conditions on some federal lands make it unlikely that a fire would burn or that a fire would threaten a nearby population.

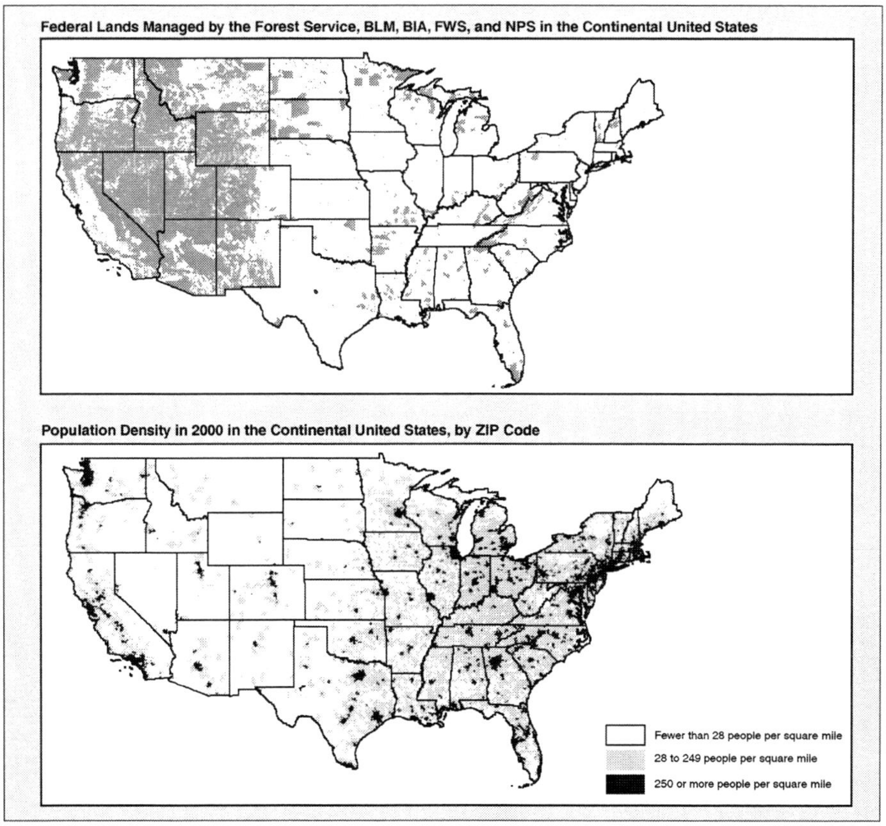

Source: GAO analysis of U.S. Censusand U.S. Geological Survey National Atlas Website data.

Note: We conducted our analysis using census data on the average population per square mile across areas defined by ZIP codes. However, especially in larger ZIP codes, there may be smaller pockets where the population density is higher or lower than the average used in our analysis. When mapping the data, we included a 1.5-mile buffer around the ZIP code areas with 28 or more people per square mile to account for the 1.5-mile buffer specified in the HFRA definition for wildland-urban interface.

Figure 8. Location of Federal Lands and Populated Areas in the Continental United States.

While many important contextual details are not visible on a national map, some can be seen at the county level. For example, in Los Angeles County-the most populous U.S. county-many of the fuel reduction treatments completed in 2005 and 2006 were adjacent to densely populated areas, as shown in figure 9, but some were miles away and in ZIP code areas with relatively low population.

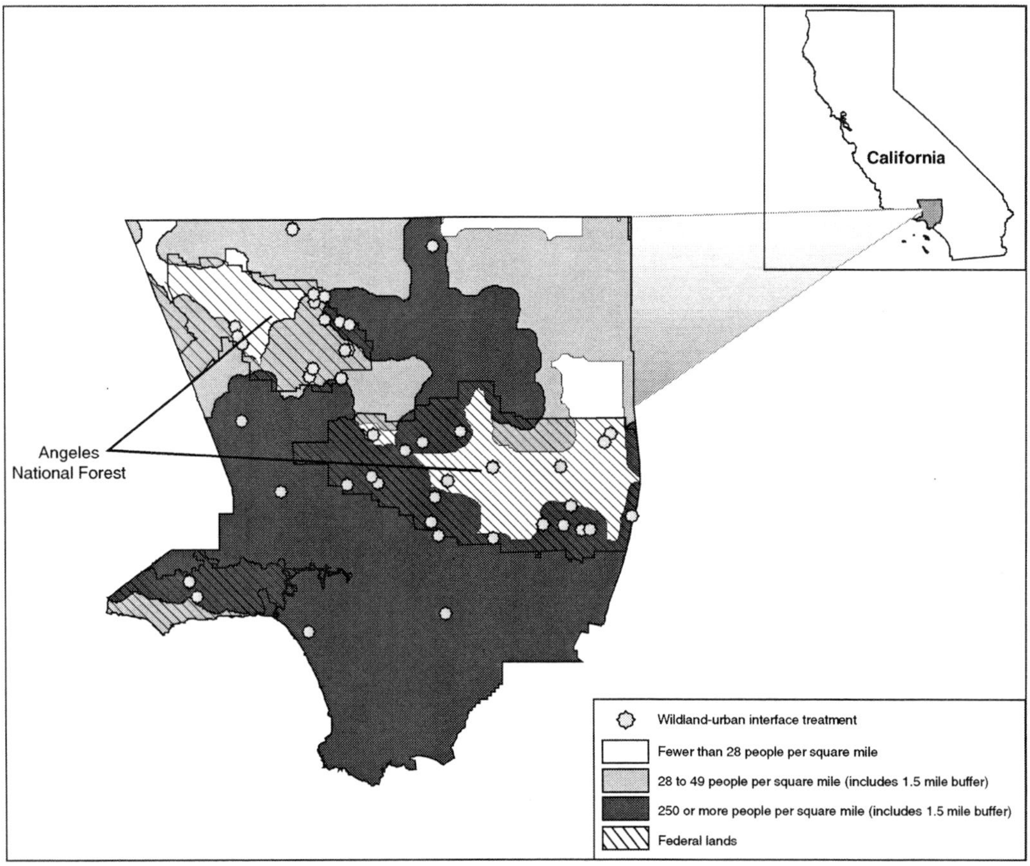

Source: GAO analysis of Forest Service, Interior, U.S. Cen sus, and U.S. Geological Survey National Atlas Web site d ata.

Note: We conducted our analysis using census data on the average population per square mile across areas defined by ZIP codes. However, especially in larger ZIP codes, there may be smaller pockets where the population density is higher or lower than the average used in our analysis.

Figure 9. Map of Los Angeles County Wildland-Urban Interface Fuel Reduction Treatments Completed in 2005 and 2006, and Population Density.

Treatments occurred in these low-density areas for several reasons. First, many of the treatments conducted in the county during that period, while not immediately adjacent to the city of Los Angeles, were on the federal land closest to the city, the Angeles National Forest, which of course is not highly populated. Also, while the average population density for the general area is low, individual communities with populations ranging from about 1,000 to 3,000 are located inside the boundaries of the forest, and the Forest Service conducted some treatments to protect them. Second, developed sites—such as campgr ounds, roads, and recreation areas— where people temporarily congregate may not be reflected on a census

map of population density. According to officials at the Angeles National Forest, human-caused wildland fires generally coincide with such areas, making it important to conduct fuel treatments around these sites. Finally, low-density areas within the forest were more feasible to treat than some areas closer to population centers because steep terrain across much of the forest—including along its southern boundary adjacent to heavily populated Los Angeles—makes it difficult and expensive to conduct fuel treatments, and, in some cases, would make treatments ineffective, according to agency officials.

In contrast to Los Angeles County, Rio Blanco County, Colorado, is a rural county with a total population of about 6,000 and an average population density throughout the county of less than 28 people per square mile. Nevertheless, BLM classified some of its fuel reduction treatments in this county as wildland-urban interface treatments. As figure 10 shows, these treatments in 2005 and 2006 were generally located far from the largest towns in the county—Meeker and Rangely—which each has a population of about 2,000.

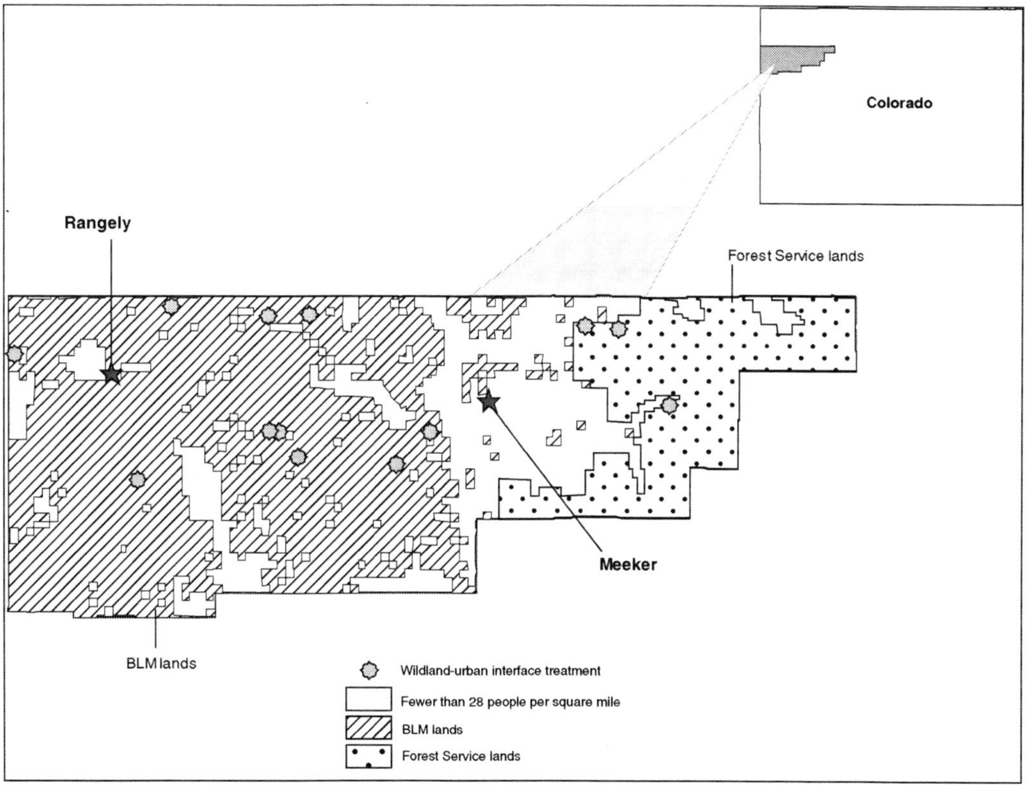

Source: GAO analysis of Forest Service, Interior, U.S. Census, and U.S. Geological Survey National Atlas Web site data.

Note: We conducted our analysis using census data on the average population per square mile across areas defined by ZIP codes. However, especially in larger ZIP codes, there may be smaller pockets where the population density is higher or lower than the average used in our analysis. In Rio Blanco County, there were no ZIP code areas with an average population of 28 or more people per square mile.

Figure 10. Map of Rio Blanco County Wildland-Urban Interface Fuel Reduction Treatments Completed in 2005 and 2006, and Population Density.

According to BLM officials, they did not conduct wildland-urban interface treatments closer to these towns because the towns are not at significant risk from wildland fire; they are surrounded in large part by rocky outcroppings and irrigated agricultural fields, where fires would not likely start, and area roads serve as fire breaks. In addition, the county had prepared a community plan identifying its highest priorities, and the federal lands surrounding the towns were not among them. Instead, many of the fuel reduction treatments the agencies did implement—including the five located southwest of Meeker—were conducted to protect energy development facilities, such as a coal mine and oil and gas wells, or power lines that service such facilities, according to BLM officials. According to the officials, they selected these projects because they were higher priority than other potential projects in the county, and because the county's community plan had identified the protection of energy development and power lines as its priorities, defining the wildland-urban interface to include the areas around such infrastructure. While these decisions may be reasonable given local priorities, it is not clear from a national perspective whether the values at risk in this case are of higher priority than the values at risk in other locations—in part because the definitions of the wildland-urban interface do not distinguish the relative importance of different values at risk, such as homes, power lines, or oil and gas wells, among others.

The Agencies Do Not Consider Treatment Effectiveness in Their Allocation Processes Because They Have No Measure for Effectiveness

Although the agencies recognize the importance of measuring the effectiveness of fuel reduction treatments—that is, how much risk reduction is achieved through a given treatment and for how long—none of the agencies considered effectiveness when allocating funds in 2007 because they have not yet developed a method for measuring it. Without understanding the potential effectiveness of fuel reduction treatments, the agencies cannot ensure that funds are allocated appropriately, because not all areas that rank high in a risk assessment can be treated with the same degree of success. For example, parts of southern California are dominated by chaparral ecosystems, which feature plants with fire-resistant roots, enabling the plants to re-sprout quickly. Some of the plants also encourage fire because their leaves are coated with a flammable resin. Although these areas of chaparral ecosystems would score high on a risk assessment—because there is a high vegetation hazard near populated areas with considerable values at risk—agency officials told us that fuel reduction treatments in chaparral may be effective for only a short time because the vegetation often grows back quickly. In addition, many of the damaging fires in southern California chaparral have been fanned by the warm, dry, and extremely powerful Santa Ana winds, making it difficult for fuel treatments to affect fire severity, according to some Forest Service officials. As a result, some of these areas, though at high risk from fire, might not be designated as high priority for fuel treatments. In general, understanding the expected effectiveness of fuel reduction treatments under different conditions can help the agencies target their funds toward treatments that will achieve the most risk reduction for a given cost. The agencies have, on multiple occasions, recognized the significance of treatment effectiveness; for example, in the 2006 10-Year Strategy Implementation Plan, the agencies identified the need to "explore the feasibility of developing measures that determine the degree and longevity of fire hazard reduction achieved by hazardous fuels treatments."

Although the agencies have not yet developed a measure of effectiveness, they have designed their allocation models to accommodate data on effectiveness in the expectation that such data will eventually become available. The Forest Service's assess effectiveness, but, because the agency does not have data on effectiveness, one of the elements serves as a placeholder—by assigning each region an identical score—and thus does not influence priority scores, while the other uses data on the total number of acres treated in each region instead. Forest Service officials acknowledged that the number of acres treated does not reveal how effective the treatments are in reducing risk, but told us they used this information because they wanted a measure that would reflect the variation in accomplishment levels from one region to the next. Interior and BLM also plan to include a measure of effectiveness in their allocation models, but Interior—like the Forest Service—currently uses total acres treated, and BLM uses data on efficiency, including total acres treated and average cost per acre, because these are the only data available. According to agency officials, it is difficult to develop a single measure of effectiveness for different geographic locations and vegetation types, because, for example, a treatment in grass might be effective for 1 year, while a treatment in some forests might be effective for 30 years. Nevertheless, as long as the agencies continue to allocate funds without knowing how effective treatments are likely to be, they cannot be sure that funds are being spent on projects that substantially reduce overall risk.

According to Forest Service research scientists, developing a measure of treatment effectiveness would require that the agencies first determine how to estimate the level of risk in a given location so they could track any changes in risk resulting from fuel treatments. For example, they could use data on fire intensity, severity, or occurrence, or some combination of these and other factors, to evaluate risk. Once agency officials determined how to estimate risk, they could use the information to measure treatment effectiveness. However, there is no consensus on how best to do so and any method would likely require considerable effort. For example, under one approach described by the researchers, available scientific studies about fuel reduction treatments in various vegetation types would be analyzed to ascertain where fuel treatments are more or less effective, and effectiveness ratings would be calculated for each vegetation type on the basis of this information. After establishing the ratings, they would collect field data to verify their initial conclusions and ratings—a costly and time-consuming exercise, according to some researchers. Such an approach would have drawbacks, however; the researchers told us that it would be difficult to establish a single rating that would apply to vegetation types under all circumstances because fuel conditions within a given vegetation type vary widely, depending, for example, on geographic location and previous fuel reduction activity. In addition, factors other than vegetation—such as terrain, weather, and soil—also influence treatment effectiveness. Consequently, some researchers have proposed alternative approaches, such as one that would consider many factors, in addition to vegetation type, to assign effectiveness ratings to individual treatment areas rather than general vegetation types. However, developing an effectiveness rating scheme using this approach—or others that incorporate numerous factors—would require significant research and analysis over a long time period, according to one researcher.

A less expensive, quicker approach outlined by another Forest Service researcher would rely on expert opinion rather than field data. Under this simplified approach, a panel of experts with knowledge about and experience in fuel reduction treatments and their effectiveness would use their professional judgment to collectively estimate the extent to which fuel treatments would be effective in each of several vegetation or fire regime condition class categories. The expert's estimates could then be used to inform decisions on allocating funds.

The Agencies Often Consider Costs, but Not Cost-Effectiveness, When Allocating Funds

The agencies also do not consider the cost-effectiveness of treatments when allocating funds, primarily because they do not have data on treatment effectiveness. Treatment costs can vary widely in different areas, from as little as $10 per acre to well over $1,000 per acre, even ranging as high as $30,000 per acre under unusual circumstances, and allocating funds wisely involves not simply targeting those acres that can be treated most cheaply, but those acres where treatments yield the most cost-effective result. While considering costs is an important step in making allocation decisions, it is equally important to consider effectiveness in conjunction with costs to avoid funding ineffective projects simply because they are cheap. However, until the agencies have data on treatment effectiveness, they will find it difficult to do so. In support of these considerations, the 2006 Cohesive Strategy emphasized the importance of reducing fuel in the most cost effective manner possible, because federal funds can support only a finite number of fuels treatments each year covering a fraction of the acres at high risk from unusually severe fires.

In practice, the agencies frequently consider costs when allocating funds and selecting projects. They sometimes give priority to projects with low per-acre costs in order to leave more funds available for other projects or to treat more acres within their budgets—an important factor for agencies trying to meet increasing acreage targets. Also, agencies sometimes give priority to low-cost treatments in areas that have previously been treated and are consequently of relatively low risk, in order to prevent them from becoming higher risk. According to agency officials, these treatments are a priority because they are a cost-effective way to maintain low-risk conditions once achieved; it is generally much cheaper to reduce fuel in areas that have recently been treated than to do so in areas that have never been treated or have not been treated for a long time. However, without knowing the effectiveness of treatments in reducing risk, agency officials may not be able to compare the relative benefits of potential projects when deciding where to invest fuel reduction funds—and, thus, may not know which projects are likely to be the most cost-effective.

In some cases, the agencies also give lower priority to treatments with very high per-acre costs—even in high-risk areas—because the expected benefit does not justify the expense. For example, the Desert National Wildlife Refuge in southern Nevada identified a mechanical thinning treatment to remove palm trees as its highest-priority fuel reduction project in 2006. The proposed project was in the wildland-urban interface and would also improve the habitat of an endangered fish, according to agency officials. However, it would have cost hundreds of thousands of dollars—nearly the entire budget for the region—and, therefore, FWS regional officials did not fund the project.

The Agencies Have Not Established Clear Guidance on the Relative Importance of Factors Used in Setting Priorities

In addition to more consistently using information on risk and developing measures of treatment- and cost-effectiveness, the agencies could improve their allocation process by clarifying the relative importance of the different factors they use in setting priorities. Without such clarification, it is not clear how agency officials are to resolve conflicts that arise between competing factors. In addition, when factors other than risk, treatment effectiveness, and cost effectiveness have considerable influence on allocation decisions, it is difficult for the agencies to ensure that funds are allocated to areas where they will most effectively reduce risk.

The agencies consider such factors in part because they are directed to do so; many of the factors they consider are tied to federal laws or congressional direction. For example, fuel reduction projects authorized under HFRA include, among others, projects on federal land in the wildland-urban interface and certain projects in areas where ecological restoration is needed because vegetation has departed significantly from its historical regime. The act requires the agencies to develop annual programs of work for federal land that give priority to authorized hazardous fuel reduction projects that provide for the protection of at-risk communities or watersheds or that implement community wildfire protection plans. Congressional committee direction has also called for the agencies to put a priority on fuel reduction work completed through mechanical treatments and projects that use biomass.

Some of the factors the agencies consider are also intended to encourage efficiency in the fuel reduction program, as well as more broadly in their land management missions. Specifically, the agencies give priority to projects that achieve benefits not only for the fuel reduction program but also for other programs such as wildlife management and watershed improvement—an approach referred to as integration among programs. Agency officials said implementing such projects is a way to leverage funds and coordinate resources. The Forest Service also emphasizes these projects because its interpretation of the President's HFI calls for a focus on integrated management, according to agency officials.

In the face of multiple directives and competing agency priorities, agency officials must balance numerous factors when allocating funds and selecting projects, as the following examples illustrate:

- *Priorities in community plans may not always align with agency-identified priorities, forcing agencies to choose between them.* According to Montana BLM officials, one community proposed a fuel reduction project in an area the officials believed was relatively low risk because it had vegetation that does not burn easily. However, the officials agreed to implement the project because they are directed to give priority to locally identified projects and because they did not want to damage their relationship with the community. Several agency officials told us that community plans did not always include federal lands or propose projects in locations where the agencies could feasibly implement a treatment. In such cases, agency officials sometimes worked with the communities to identify project locations agreeable to all, while other times they implemented agency-identified projects instead of those identified in the plans.
- *Direction to give priority to high-risk areas may also conflict with the agencies'* Several agency officials told us that they sometimes implemented lower-priority

projects with low unit costs because they felt pressure to meet acreage targets. In some cases, these projects, although low priority for fuel reduction purposes, were a high priority for other resource programs or achieved other management objectives.

- *Direction to give priority to areas in the wildland-urban interface may conflict with other agency priorities.* For example, NPS officials told us that giving priority to fuel reduction treatments at the interface conflicted with the agency's and processes, which would call for giving priority to treatments in undeveloped areas.
- *Desire for stable funding and staff levels may make officials reluctant to shift funds on the basis of risk assessments.* When allocating funds, the agencies frequently emphasized the importance of maintaining stable funding levels and minimizing disruptions to staff, which can conflict with the direction to emphasize high-risk areas. According to agency officials, stable allocations to regions and field units are needed to ensure predictability and enable regional and field staff to plan ahead. In addition, a minimum level of funding is needed to maintain the workforce and infrastructure required to support viable fuel reduction programs in regions and field units. Several agency officials told us they were reluctant to shift funding on the basis of risk assessments because doing so could require staff to relocate—potentially multiple times—and the officials wanted to avoid uprooting staff.

Agency guidance offers little in the way of clarification for staff confronted with numerous, conflicting priorities, as the multiplicity of priorities in the

Forest Service and Interior's strategy, the Forest Service and Interior outline a set of national fuel treatment priorities but do not establish a hierarchy of their relative importance. Among the treatment priorities are areas in the wildland-urban interface as well as some areas outside the interface in condition classes 2 or 3. In addition, some areas in condition class 1—the only remaining condition class—are to be gi ven equal priority, according to the strategy. Similarly, the strategy calls for priority to be given to mechanical treatments where appropriate, but also to prescribed burns where appropriate. After providing a list of priority criteria, the strategy declares that the more criteria a fuel reduction project meets, the higher its priority should be for funding. However, it also acknowledges that, in exercising management discretion, the agencies may need to make exceptions to the process described for ranking and selecting projects.

Agencies' Allocation Processes Are Not Always Systematic

Although the agencies are working to develop and implement models that will allow them to allocate funds more systematically, such systematic approaches are not used by all agencies or at all levels within the agencies. By allocating funds using a systematic process—one that is methodical, based on established criteria, and applied consistently—the agencies can better ensure that they uniformly consider all relevant criteria and appropriately apply these criteria in all decisions.

In particular, when agency officials rely primarily on professional judgment and negotiation to allocate funds, they do not always follow a step-by-step approach or consistently apply a predetermined set of criteria. We recognize that agency decision makers—particularly those who have served in the same location for many years—often have

detailed knowledge about on-the-ground conditions and a thorough understanding of fuel reduction needs. Nevertheless, without using a systematic approach, even knowledgeable and well-meaning decision makers may be more susceptible to influences that are not intended to be part of the decisions, as illustrated by the following examples:

- According to several agency officials, they face considerable pressure to meet acreage targets. Under these circumstances, and with no predetermined set of criteria in an allocation process, targets could have more influence than intended. That is, agency officials might fund lower priority projects in order to treat more acres.
- In NPS's Southeast full-time fuel reduction staff has considerable influence on allocations, even though it is not officially a factor in the allocation process. Few parks in this region have full-time staff devoted to fuel reduction, and parks without such staff request and receive much less fuel reduction funding than do the parks with dedicated staff—potentially because there are fewer staff to perform the work necessary to identify fuel reduction needs and request funds. Consequently, according to agency officials, it is difficult to ensure that all of the highest-priority areas for fuel reduction across the region are identified and targeted for funding because some high-priority areas may not be identified if they are located in parks with fewer staff. NPS officials in another region expressed a similar concern, stating that the agency needs to shift fuel reduction funds within the region to direct them to high-priority locations and acknowledging that it cannot do so without also shifting personnel to high-priority locations.

Moving toward more systematic allocation processes also enhances transparency and accountability. In many of the locations we visited, the agency offices that relied primarily on professional judgment to allocate fuel reduction funds and select projects did not document the rationale for their decisions. As a result, the processes were not transparent, and neither agency officials nor others—including Congress and the public— could understand the rationale behind the decisions or have confidence that the resulting allocations were directed to the highest-priority areas for reducing risk to communities and the environment. For example, officials in BLM's Oregon/Washington state office used their professional judgement to determine allocations to its 10 district offices. Under this process in 2007, BLM's Medford $9 million—over 7 times the average allocation received by the other nine district offices that year. While this disparity may be appropriate, without a transparent process it is difficult to determine the extent to which the allocation reflects agency priorities for reducing risk to communities and the environment, rather than other factors. The agencies themselves have emphasized the importance of transparency and accountability; for example, the 10-Year Strategy Implementation Plan states that the agencies should "strive for maximum transparency in the decision-making process."

CONCLUSIONS

Our nation's wildland fire problem has been decades in the making and wildland will not be solved quickly. Nevertheless, with careful choices about where to spend their limited fuel reduction dollars, federal agencies can meaningfully, if incrementally, reduce the risks faced

by communities and the environment. Doing so will require the agencies to continue moving away from allocation by tradition to allocation by priority. Toward this end, the agencies could improve their current approaches in three key areas.

First, the agencies would benefit from routinely using an allocation process that is systematic, and that is common to all the agencies. A systematic process can help ensure that the agencies apply their allocation and project selection criteria consistently, and can help interested parties outside of the process—Congress, local communities, and other entities—understand the rationale for the funding and project selection decisions that are made. While the models that some of the agencies are developing represent substantial steps forward in this regard and will affect larger portions of funding allocations over time, not all of the agencies have models, and none consistently uses models at the national, regional, and local levels. Further, the models, even where used, often exert only a small influence on allocation decisions, partly because the agencies do not yet have full confidence in the models' base decisions mainly on historical funding patterns and professional judgment. We recognize that professional judgment will always have a role in the allocation process to account for difficult-to-quantify factors, such as local priorities or political considerations. However, the agencies and the public are best served if a systematic process, such as a model, serves as the foundation for allocation decisions, and professional judgment plays a supporting, rather than a lead, role. Also, given that wildland fire is a nationwide problem that does not respect administrative boundaries, the agencies would do well to develop and use a common process for allocating fuel reduction funds—as Congress has called for—that can be customized to accommodate differences in scale, type of ecosystem, agency mission, and other criteria.

Second, the agencies could improve the information they use to make allocation decisions. Because the agencies do not always use risk assessments and currently lack data on treatment effectiveness, they often make allocation decisions without knowing, on a broad scale, where the acres at highest risk are located, which treatments are most effective at reducing risk, and which areas respond best to treatment. To improve their allocation decisions, they should continue, over the long term, to develop and use information on risk and treatment effectiveness. The agencies can then use this information, in concert with cost information, to effectively assess tradeoffs among potential treatments and identify the most cost-effective investments.

Finally, the agencies could strengthen their allocation processes by sorting through the numerous prioritization factors that have accumulated over the years and establishing a hierarchy for considering them. Without such a hierarchy, the exercise of setting priorities can be frustrating—or even meaningless—because virtually any project can qualify as high priority. While we recognize that, in some cases, the agencies are bound by law or congressional direction to give priority to certain factors, we believe there may remain enough room within those constraints not only to establish a hierarchy of factors, but also to clarify the relative importance of categories within some factors—in particular, various categories of wildland-urban interface. We do not advocate prioritization by census— simply directing fuel reduction funds to areas with the highest populations—but neither do we believe that the agencies or the public are well-served by the broad definitions of the interface currently used. However, if the agencies determine, through further analysis, that laws or congressional direction create conflicts prohibiting them from implementing a consistent, systematic approach that distinguishes the relative importance of various priorities, they should so inform Congress and seek additional clarification.

It will not be easy to carry out these tasks. As they work to improve their processes, the agencies will need to devote considerable effort to developing measures and collecting data on risk and effectiveness and considerable thought to balancing this information against the many goals of the fuel reduction program—all in a way that yields transparent results. And once these steps are carried out, the agencies face perhaps an even more difficult decision: how best to redirect fuel reduction funds in a way that improves the agencies'use of their limited funds despite the potentially disruptive consequences for individual field units or nearby communities. Our findings suggest that the agencies are increasingly mindful of the merits of such an approach and that their recent actions have begun to lay the necessary groundwork. Nevertheless, many challenges remain, and a difficult road lies ahead.

APPENDIX I: OBJECTIVES, SCOPE, AND METHODOLOGY

We were asked to (1) identify the processes the Forest Service, the Department of the Interior (Interior), and Interior's agencies –the Bureau of Indian Affairs (BIA), Bureau of Land Management (BLM), Fish and Wildlife Service (FWS), and National Park Service (NPS)—use to allocate fuel reduction funds and select projects for implementation, including the factors that influence these processes; and (2) determine how, if at all, the agencies could improve these processes to better ensure they contribute to their goal of effectively reducing the risk of wildland fire to communities and the environment. We focused our review primarily on the Forest Service and BLM because these two agencies accounted for about 80 percent of the fuel reduction funds appropriated by Congress for 2005, 2006, and 2007, although we collected information on the other three agencies as well.[1] We focused our review on fuel reduction work funded through congressional fuel reduction appropriations; therefore, fuel reduction work funded by other agency programs or outside organizations is outside the scope of this review. To gain an understanding of outside perspectives on the agencies' nonfederal parties, including representatives from the National Association of State Foresters, The Nature Conservancy, the Western Governors' Association, and the Wilderness Society.

Fuel Reduction Funding Allocation and Project Selection Processes

To learn how the agencies allocate fuel reduction funds and select projects, and to identify the factors that influence these processes, we first obtained and reviewed documents on policies and procedures governing the fuel reduction program. These included applicable laws, administrative initiatives, congressional committee reports, and interagency agreements, as well as guidance for fuel reduction from the departments, agency headquarters, and regional offices.[2] We also obtained and analyzed agency data on funding allocations.

To learn about the processes used to allocate fuel reduction funds at the national level, we met with agency officials from the Forest Service and Interior at their Washington, D.C., headquarters, and with officials from all five agencies at the National Interagency Fire Center in Boise, Idaho. We also met with agency researchers and modeling experts to better understand the data used in the national models currently under development by the Forest

Service, Interior, and BLM. We did not, however, assess the accuracy or technical soundness of these models.

At the regional and state levels, we used a structured interview guide to speak, in person or by telephone, with officials in all Forest Service regional and BLM state offices, as well as with officials in selected BIA, FWS, and NPS regional offices.

The structured interview guide included questions about the processes used to allocate fuel reduction funds, the factors that influence those processes, the extent and nature of regional guidance provided to field units, and the amount of oversight on the part of the regional offices. Because developing and administering a structured interview guide may introduce errors—caused by the way a particular question is interpreted, for example—we included steps in the development and administration of the interview guide to minimize such errors. We pretested the guide at several locations and modified it to reflect questions and comments we received. We also visited a number of the agencies' regional offices to obtain a greater understanding of the funding allocation processes in those regions. We selected regional offices that collectively received a substantial portion of their agency's fuel reduction funds and represented diversity with respect to fuel reduction funding levels, fuel reduction acreage accomplishments, predominant vegetation type, and geographic location. These selection criteria are shown in table 5.

To learn about the project selection processes used by local units, we selected a nonprobability sample of 20 local units in eight states to interview.[3] The sample included 8 national forests, 5 BLM district or field offices, 2 BIA agencies, 2 national wildlife refuges, and 3 national parks. Table 6 lists the units we visited. The local units selected for interviews represented diversity with respect to geographic location and predominant vegetation type. In addition, we selected units that represented diversity with respect to their proximity to communities and development, including units that were located in counties that were predominantly rural or urban.

During all of these visits, we collected documents and interviewed staff; during some of these visits, we also observed fuel reduction treatments. Because we conducted in-depth analyses of only a few selected units, we cannot generalize our findings beyond the local units and officials we contacted.

Potential Improvements to Agency Processes to Better Ensure They Contribute to Reducing Risk

To identify potential improvements to the agencies' processes for allocating fuel reduction funds and selecting treatments, we analyzed the information we collected through our site visits, structured interviews, agency documentation, and discussions with other agency officials. To identify the overall goals of the fuel reduction program, and the extent to which earlier assessments of the program identified shortcomings in the agencies' ability to meet these goals, we also evaluated (1) agency policy documents, including strategy documents, program guidance, and related documents discussing the program's objectives: (2) legislative direction associated with the fuel reduction program, including laws, congressional committee report language, and other direction; and (3) previous reviews of the fuel reduction program by GAO, the Inspectors General, and others.

Table 5. Regional Offices GAO Visited

Agency	Region/state office	Fuel reduction funding level[a]		Acres treated[b]		Predominant vegetation type			Geographic location		
		Greater than average	Less than average	Greater than average	Less than average	Forest	Grass	Shrub	West	Central	East
Forest Service	Northern		x		x	x				x	
	Pacific Northwest	x			x	x			x		
	Pacific Southwest	x			x	x			x		
	Rocky Mountain	x			x	x				x	
	Southern	x		x		x					x
BIA	Northwest	x		x		x	x		x		
	Pacific (by phone)		x		x			x	x		
	Rocky Mountain		x		x	x	x			x	
	Southwest	x		x			x			x	
BLM	California		x		x		x	x	x		
	Colorado		x		x			x		x	
	Idaho	x		x			x		x		
	Montana		x		x		x			x	
	Oregon/Washington	x		x		x	x		x		
FWS	California-Nevada[c]		x		x			x	x		
	Mountain-Prairie	x			x		x			x	
	Southeast	x		x		x					x
NPS	Intermountain	x		x		d	d	d		x	
	Pacific West	x		x		d	d	d	x		
	Southeast	x		x		x					x

Source: GAO analysis of Forest Service and Interior data.

[a]"Greater than average" refers to regions thatreceived more than the average funding amount received by that agency's regions than average" refers to regions that receivedin 2007, and "less less than the average amount in 2007.

[b]"Greater than average" refers to regions that treated more than the average acres treated by that agency's regions in 2006, and "less than average" acres treated in 2006.

[c]FWS's ffice is officially part of the Pacific region, but manages its ownCalifornia-Nevada Operations o fuel reduction program.

[d]These regions each cover several states and have a large variety of vegetation; therefore, no one vegetation type is predominant, according to agency officials.

Table 6. Field Units GAO Visited

Agency and unit	State	Predominant vegetation type Forest	Grass	Shrub
Forest Service				
Angeles National Forest	California			x
Arapaho-Roosevelt National Forests and Pawnee National Grassland	Colorado	x		
Bitterroot National Forest[a]	Montana and Idaho	x		
Boise National Forest	Idaho	x		
Chattahoochee-Oconee National Forests	Georgia	x		
Medicine Bow-Routt National Forests and Thunder Basin National Grassland	Wyoming and Colorado	x		
National Forests in Florida (Ocala National Forest)	Florida	x		
San Bernardino National Forest	California	x		x
BIA				
Crow Agency	Montana		x	
Zuni Agency	New Mexico	x		
BLM				
Albuquerque District Office	New Mexico	x		
Billings Field Office	Montana		x	
Little Snake Field Office	Colorado		x	x
Twin Falls District Office	Idaho		x	
White River Field Office	Colorado	x		x
FWS				
Merritt Island National Wildlife Refuge	Florida			x
Rocky Mountain Arsenal National Wildlife Refuge	Colorado		x	
NPS				
Cape Canaveral National Seashore	Florida	x		
Rocky Mountain National Park	Colorado	x		
Sequoia and Kings Canyon National Parks[a]	California	x		

Source: GAO analysis of Forest Service and Interior data.
[a]We met with officials from these field units at off-site locations, in order to facilitate cost-effective travel logistics.

In our interviews with agency officials, we asked about the factors they considered when allocating funds and selecting projects—including the influence of specific factors, such as acreage targets and risk assessments—as well as factors that prevented high-priority work from being accomplished. We also asked about regional and local definitions of the wildland-urban interface. We assigned the allocation processes used by the agencies' two categories: quantitative or judgmental. We also verified the factors used in allocation processes with agency officials.

To determine the extent to which the locations of wildland-urban interface treatments, population centers, and federal lands coincided, we analyzed fuel reduction data from a Forest Service and Interior database—the National Fire Plan Operations and Reporting System (NFPORS)—as well as population data from the U.S. Census Bureau, and federal lands data from the U.S. Geological Survey's (NationalAtlas.gov). Using the agency data on fuel reduction treatments, we used geographic information system (GIS) tools to map the location and size of wildland-urban interface treatments completed in 2005 and 2006. We also applied GIS tools to Census data to map population density in three categories: (1) fewer than 28 people per square mile, on average; (2) 28 to 249 people per square mile, on average; and (3) 250 or more people per square mile, on average.[4] We used these categories because they reflect the definition of wildland-urban interface published in the January 4, 2001 Federal Register and used by the agencies.[5] We also mapped the location of federal lands using the data from the U.S. Geological Survey. In addition, we created maps of two counties—one urban and one rural—showing the lo cations of wildland-urban interface treatments completed in 2005 and 2006, population density, and federal lands. We also contacted officials in those two counties to discuss the location of specific wildland-urban interface projects and their rationale for selecting those projects.

To determine the reliability of the agencies' fuel reduction data, we reviewed related documentation, such as the NFPORS database manual; interviewed knowledgeable agency officials, including database administrators; discussed data input and verification procedures with regional and field staff; and conducted electronic data testing. We found that these fuel reduction data were sufficiently reliable for the purposes of this review. We obtained the federal lands data prepared by NationalAtlas.gov and reviewed the documentation provided on the limitations of the file. From this review, we determined that the federal lands data were sufficiently reliable for our purposes. To measure population density, we used Census ZIP Code Tabulation Area data from the 2000 U.S. Census and the geographic boundary for those areas. We reviewed documentation provided on the limitations of these files and compared their consistency with other Census sources. From this review, we determined that the population density data were sufficiently reliable for our purposes.

We conducted our work from August 2006 to September 2007 in accordance with generally accepted government auditing standards.

APPENDIX II: FOREST SERVICE AND INTERIOR FUEL REDUCTION FUNDING ALLOCATIONS, FISCAL YEARS 2005, 2006, AND 2007

This appendix provides information on fuel reduction funding appropriations and allocations to the Forest Service, the Department of the Interior (Interior) and its agencies, and their regions for 2005, 2006, and 2007.[1] Interior allocates separate fuel reduction funds to its four agencies for treatments within and outside of the wildland-urban interface (WUI), while the Forest Service allocates one single source of fuel funding to its regions. Therefore, information on allocation amounts to WUI and non-WUI areas are included for Interior but not for the Forest Service. Table 7 provides total appropriations and allocations to the Forest Service and Interior agencies—Bureau of Indian Affairs (BIA), Bureau of Land Management (BLM), National Park Service (NPS), and Fish and Wildlife Service (FWS)—for 2005, 2006, and 2007. As shown in table 7 and figure 11, of the approximately $500 million appropriated to the Forest Service and Interior for fuel reduction in 2007, the Forest Service received about 61 percent of the total; BLM received about 19 percent of the total; and the remaining 20 percent was allocated to BIA, NPS, and FWS.

Table 7. Total Appropriations to Forest Service, and Allocations to Interior Agencies, Fiscal Years 2005, 2006, and 2007

Agency	Total allocation	Percentage of total allocation
BLM		
2005	$91,386,000	18.8
2006	96,299,000	19.9
2007	93,389,000	18.8
BIA		
2005	42,488,000	8.7
2006	43,237,000	8.9
2007	40,664,000	8.2
NPS		
2005	33,040,000	6.8
2006	33,299,000	6.9
2007	31,396,000	6.3
FWS		
2005	27,527,000	5.7
2006	32,162,000	6.6
2007	30,666,000	6.2
Subtotal—Interior agencies		
2005	**194,441,000**	**39.9**
2006	**204,997,000**	**42.3**
2007	**196,115,000**	**39.4**
Forest Service[a]		
2005	292,389,000	60.1
2006	280,119,000	57.7
2007	301,258,000	60.6
Total		
2005	**486,830,000**	**100**
2006	**485,116,000**	**100**
2007	**497,373,000**	**100**

Source: GAO analysis of Forest Service and Interior data.

Notes: Interior allocated additional amounts of $6,968,000 in 2005; $5,115,000 in 2006; and $3,672,000 in 2007 to the Office of Wildland Fire Coordination, which is responsible for the coordination, integration, and oversight of wildland fire management programs within Interior. Total allocations do not include carryover from the previous fiscal year. Numbers may not total due to rounding.

[a]Forest Service figures represent appropriations.

Table 8 shows the Forest Service's headquarters for 2005, 2006, and 2007. In 2007, of the total funding allocated to the Forest Service for fuel reduction, about 68 percent was allocated to the regions. Approximately 32 percent was allocated to the Forest Service's that are used for expenses not charged to a single program, including indirect, support, and common services charges. The Pacific Southwest region received the most funding for 2005, 2006, and 2007; the Southwestern region received the second-most funding during that time period.

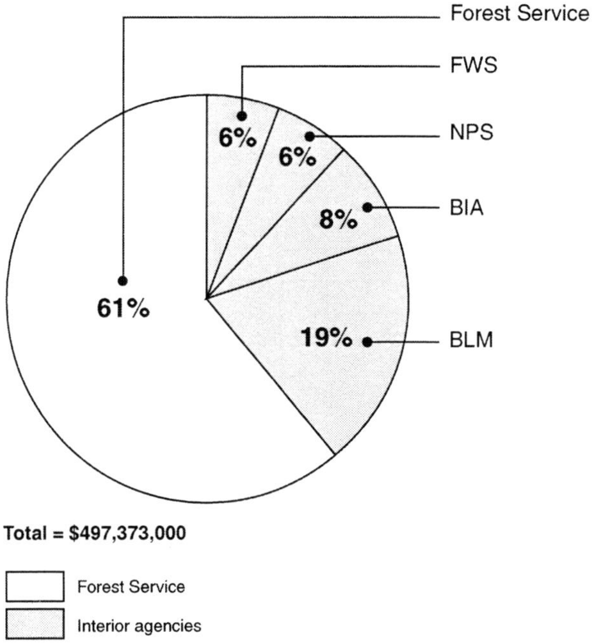

Source: GAO analysis of Forest Service and Interior data.

Notes: Interior allocated an additional $3,672,000 to the Office of Wildland Fire Coordination. Total allocations do not include carryover from the previous fiscal year.

Figure 11. Agency Funding Levels as a Percentage of Total Fuel Reduction Funding, Fiscal Year 2007.

Table 8. Forest Service Allocations to Regions and Headquarters, Fiscal Years 2005, 2006, and 2007

Region	Total allocation	Percentage of Forest total allocation
Pacific Southwest		
2005	$66,656,000	22.8
2006	41,944,000	15.0
2007	43,737,000	14.5
Southwestern		
2005	30,638,000	10.5
2006	36,891,000	13.2
2007	37,341,000	12.4
Southern		
2005	25,478,000	8.7
2006	26,368,000	9.4
2007	29,092,000	9.7
Pacific Northwest		
2005	24,622,000	8.4
2006	23,179,000	8.3
2007	25,794,000	8.6

Table 8. (Continued)

Region	Total allocation	Percentage of Forest total allocation
Rocky Mountain		
2005	21,032,000	7.2
2006	23,706,000	8.5
2007	25,445,000	8.4
Intermountain		
2005	13,673,000	4.7
2006	15,881,000	5.7
2007	16,165,000	5.4
Northern		
2005	11,875,000	4.1
2006	12,006,000	4.3
2007	15,782,000	5.2
Eastern		
2005	8,633,000	3.0
2006	8,631,000	3.1
2007	9,718,000	3.2
Alaska		
2005	1,834,000	0.6
2006	853,000	0.3
2007	805,000	0.3
Subtotal, regions		
2005	**204,441,000**	**69.9**
2006	**189,459,000**	**67.6**
2007	**203,879,000**	**67.7**
Headquarters, Research stations, and cost pools		
2005	87,948,000	30.1
2006	90,659,000	32.4
2007	97,379,000	32.3
Total		
2005	**292,389,000**	**100**
2006	**280,119,000**	**100**
2007	**301,258,000**	**100**

Source: GAO analysis of Forest Service data.

Notes: Total allocations do not include carryover from the previous fiscal year. Numbers may not total due to rounding.

Table 9 shows Interior's including WUI and non-WUI allocations—for 2005, 2006, and 2007. In 2007, about 65 percent of Interior's 35 percent was to non-WUI areas. In 2007, BLM received the largest percentage of Interior's 48 percent.

Table 10 shows BLM's for 2005, 2006, and 2007. In 2007, about 71 percent of BLM's total fuel reduction funding was allocated to WUI areas, while about 29 percent was allocated to non-WUI areas. In 2005, 2006, and 2007, the Oregon/ Washington state office received the most funding, followed by the Idaho and Utah state offices. These three offices accounted for about 50 percent of BLM's total annual fuel reduction allocation.

Table 9. Interior Allocations to BLM, BIA, FWS, and NPS, Including WUI and Non-WUI Allocations, Fiscal Years 2005, 2006, and 2007

Agency	Total allocation	Percentage of Interior's total allocation to the agencies	Total WUI allocation	Total non-WUI allocation
BLM				
2005	$91,386,000	47.0	$64,437,000	$26,949,000
2006	96,299,000	47.0	66,245,000	30,054,000
2007	93,389,000	47.6	66,590,000	26,799,000
BIA				
2005	42,488,000	21.9	27,299,000	15,189,000
2006	43,237,000	21.1	27,494,000	15,743,000
2007	40,664,000	20.7	26,681,000	13,983,000
NPS				
2005	33,040,000	17.0	15,320,000	17,720,000
2006	33,299,000	16.2	14,948,000	18,351,000
2007	31,396,000	16.0	14,583,000	16,813,000
FWS				
2005	27,527,000	14.2	15,583,000	11,944,000
2006	32,162,000	15.7	19,772,000	12,390,000
2007	30,666,000	15.6	18,922,000	11,744,000
Total—Interior agencies				
2005	**$194,441,000**	**100**	**$122,639,000**	**$71,802,000**
2006	**$204,997,000**	**100**	**$128,459,000**	**$76,538,000**
2007	**$196,115,000**	**100**	**$126,776,000**	**$69,339,000**

Source: GAO analysis of Interior data.

Notes: Interior allocated an additional $6,968,000 in 2005, $5,115,000 in 2006, and $3,672,000 in 2007 to the Office of Wildland Fire Coordination. Total allocations do not include carryover from the previous fiscal year.

Table 10. BLM Allocations to State Offices and Headquarters, Fiscal Years 2005, 2006, and 2007

State office	Total allocation	Percentage of BLM's total allocation	Total WUI allocation	Total non-WUI allocation
Oregon/Washington				
2005	$26,177,000	27.6	$19,027,000	$7,150,000
2006	24,596,000	24.1	17,966,000	6,630,000
2007	24,878,000	24.8	18,542,000	6,336,000
Idaho				
2005	14,536,000	15.3	10,130,000	4,406,000
2006	14,787,000	14.5	10,033,000	4,754,000
2007	14,598,000	14.6	10,338,000	4,260,000
Utah				
2005	8,557,000	9.0	5,479,000	3,078,000
2006	7,968,000	7.8	5,225,000	2,743,000
2007	10,078,000	10.1	6,164,000	3,914,000
California				
2005	7,257,000	7.6	6,096,000	1,161,000
2006	6,364,000	6.2	5,382,000	982,000
2007	7,322,000	7.3	6,294,000	1,028,000

Table 10. (Continued)

State office	Total allocation	Percentage of BLM's total allocation	Total WUI allocation	Total non-WUI allocation
Nevada				
2005	6,663,000	7.0	4,572,000	2,091,000
2006	5,794,000	5.7	3,881,000	1,913,000
2007	6,414,000	6.4	4,317,000	2,097,000
Colorado				
2005	6,480,000	6.8	4,891,000	1,589,000
2006	6,068,000	5.9	4,589,000	1,479,000
2007	6,843,000	6.8	5,285,000	1,558,000
New Mexico				
2005	5,676,000	6.0	2,930,000	2,746,000
2006	6,167,000	6.0	3,347,000	2,820,000
2007	6,412,000	6.4	3,630,000	2,782,000
Montana				
2005	5,338,000	5.6	4,248,000	1,090,000
2006	4,871,000	4.8	4,000,000	871,000
2007	5,461,000	5.5	4,366,000	1,095,000
Arizona				
2005	4,219,000	4.4	2,509,000	1,710,000
2006	3,787,000	3.7	2,396,000	1,391,000
2007	4,355,000	4.3	2,608,000	1,747,000
Wyoming				
2005	3,143,000	3.3	1,830,000	1,313,000
2006	2,898,000	2.8	1,786,000	1,112,000
2007	3,684,000	3.7	2,185,000	1,499,000
Alaska				
2005	785,000	0.8	365,000	420,000
2006	1,044,000	1.0	502,000	542,000
2007	1,556,000	1.6	786,000	770,000
Eastern states				
2005	98,000	0.1	78,000	20,000
2006	83,000	0.1	58,000	25,000
2007	126,000	0.1	90,000	36,000
Subtotal, state offices				
2005	**88,929,000**	**93.7**	**62,155,000**	**26,774,000**
2006	**84,427,000**	**82.6**	**59,165,000**	**25,262,000**
2007	**91,727,000**	**91.5**	**64,605,000**	**27,122,000**
Headquarters				
2005	5,979,000	6.3	4,527,000	1,452,000
2006	17,822,000	17.4	12,029,000	5,793,000
2007	8,473,000	8.5	6,315,000	2,158,000
Total				
2005	**94,908,000**	**100**	**66,682,000**	**28,226,000**
2006	**102,249,000**	**100**	**71,194,000**	**31,055,000**
2007	**100,200,000**	**100**	**70,920,000**	**29,280,000**

Source: GAO analysis of BLM data.

Notes: Total allocations include the allocation for the current year plus carryover from the previous fiscal year. Numbers may not total due to rounding.

Table 11 shows BIA's allocations for 2005,2006 and 2007 to its 12 regions and the National Interagency Fire Center. In 2007, about 67 percent of BIA's total fuel reduction funding was allocated to WUI areas, while about 33 percent was allocated to non-WUI areas. In 2005, 2006, and 2007, the Northwest region received the most funding of the BIA regions, followed by the Southwest region. These two regions accounted for about 50 percent of BIA's total fuel reduction allocation in 2007.

Table 11. BIA Allocations to Regions and the National Interagency Fire Center, Fiscal Years 2005, 2006, and 2007

Region	Total allocation	Percentage of BIA's total allocation	Total WUI allocation	Total non-WUI allocation
Northwest				
2005	$12,204,855	28.1	$7,823,617	$4,381,238
2006	11,387,925	26.4	8,012,751	3,375,174
2007	11,835,643	29.6	8,465,745	3,369,898
Southwest				
2005	7,651,861	17.6	3,518,105	4,133,756
2006	9,175,694	21.3	4,115,843	5,059,851
2007	8,366,522	21.0	5,128,157	3,238,365
Western				
2005	4,497,336	10.4	2,912,773	1,584,563
2006	4,020,742	9.3	2,218,682	1,802,060
2007	3,366,120	8.4	2,287,498	1,078,622
Pacific				
2005	3,846,125	8.9	2,863,206	982,919
2006	3,096,619	7.2	2,546,681	549,938
2007	2,401,976	6.0	1,453,604	948,372
Great Plains				
2005	2,464,878	5.7	1,461,855	1,003,023
2006	3,196,997	7.4	1,742,168	1,454,829
2007	2,281,299	5.7	1,061,894	1,219,405
Alaska				
2005	2,417,117	5.6	2,247,117	170,000
2006	2,256,079	5.2	2,163,458	92,621
2007	1,780,638	4.5	1,538,479	242,159
Midwest				
2005	1,876,013	4.3	1,280,813	595,200
2006	2,736,830	6.3	1,988,630	748,200
2007	2,913,975	7.3	2,497,657	416,318
Rocky Mountain				
2005	1,699,576	3.9	1,004,728	694,848
2006	1,769,815	4.1	1,015,550	754,265
2007	1,795,609	4.5	1,066,157	729,452
Navajo				
2005	1,258,919	2.9	342,421	916,498
2006	1,335,559	3.1	443,588	891,971
2007	933,183	2.3	357,126	576,057

Table 11. (Continued)

Region	Total allocation	Percentage of BIA's total allocation	Total WUI allocation	Total non-WUI allocation
Southern Plains				
2005	572,247	1.3	27,552	544,695
2006	568,471	1.3	30,174	538,297
2007	489,663	1.2	21,867	467,796
Eastern				
2005	503,175	1.2	425,375	77,800
2006	215,825	0.5	115,732	100,093
2007	433,609	1.1	272,952	160,657
Eastern Oklahoma				
2005	284,557	0.7	216,900	67,657
2006	224,465	0.5	189,709	34,756
2007	240,074	0.6	164,172	75,902
Subtotal, regions				
2005	**39,276,659**	**90.4**	**24,124,462**	**15,152,197**
2006	**39,985,021**	**92.7**	**24,582,966**	**15,402,055**
2007	**36,838,311**	**92.3**	**24,315,308**	**12,523,003**
National Interagency Fire Center				
2005	4,172,185	9.6	3,621,793	550,392
2006	3,156,566	7.3	2,720,552	436,014
2007	3,092,337	7.7	2,608,684	483,653
Total				
2005	**43,448,844**	**100**	**27,746,255**	**15,702,589**
2006	**43,141,587**	**100**	**27,303,518**	**15,838,069**
2007	**39,930,648**	**100**	**26,923,992**	**13,006,656**

Source: GAO analysis of BIA data.

Notes: Total allocations include the allocation for the current year plus carryover from the previous fiscal year. Numbers may not total due to rounding.

Table 12 shows NPS's allocations to its seven regions and the Washinton Office for 2005, 2006, and 2007. In 2007, about 46 percent of NPS's total fuel reduction funding was allocated to WUI areas, while about 54 percent was allocated to non-WUI areas. NPS was the only Interior agency that allocated more funds to non-WUI areas than to WUI areas. In 2005, 2006, and 2007, the Pacific West region received the most funding; followed by the Intermountain region. These two regions accounted for about 60 percent of NPS's total annual fuel reduction allocation in 2007.

Table 12. NPS Allocations to Regions and the Washington Office, Fiscal Years 2005, 2006, and 2007

Region	Total allocation	Percentage of NPS's total allocation	Total WUI allocation	Total non-WUI allocation
Pacific West				
2005	$12,550,296	35.5	$7,317,240	$5,233,056
2006	10,478,930	32.4	6,231,822	4,247,108
2007	10,693,592	31.6	6,401,068	4,292,524

Table 12. (Continued)

Region	Total allocation	Percentage of NPS's total allocation	Total WUI allocation	Total non-WUI allocation
Intermountain				
2005	10,070,518	28.5	4,268,238	5,802,280
2006	8,578,937	26.5	3,138,962	5,439,975
2007	9,398,600	27.8	4,006,559	5,392,041
Southeast				
2005	4,279,340	12.1	1,973,600	2,305,740
2006	4,047,002	12.5	1,648,782	2,398,220
2007	4,604,308	13.6	1,843,531	2,760,777
Midwest				
2005	3,104,724	8.8	627,690	2,477,034
2006	3,341,288	10.3	634,129	2,707,159
2007	3,469,731	10.3	644,967	2,824,764
Northeast				
2005	898,084	2.5	521,020	377,064
2006	879,070	2.7	406,936	472,134
2007	1,201,497	3.6	757,338	444,159
Alaska				
2005	560,582	1.6	0	560,582
2006	813,140	2.5	0	813,140
2007	739,037	2.2	0	739,037
National Capital				
2005	142,043	0.4	122,320	19,723
2006	99,851	0.3	99,851	0
2007	103,631	0.3	103,631	0
Subtotal, regions				
2005	**31,605,587**	**89.5**	**14,830,108**	**16,775,479**
2006	**28,238,218**	**87.4**	**12,160,482**	**16,077,736**
2007	**30,210,396**	**89.3**	**13,757,094**	**16,453,302**
Washington Office				
2005	3,705,107	10.5	1,572,455	2,132,652
2006	4,080,417	12.6	1,904,137	2,176,280
2007	3,605,604	10.7	1,804,906	1,800,698
Total				
2005	**35,310,694**	**100**	**$16,402,563**	**18,908,131**
2006	**32,318,635**	**100**	**14,064,619**	**18,254,016**
2007	**33,816,000**	**100**	**15,562,000**	**18,254,000**

Source: GAO analysis of NPS data.

Notes: Total allocations include the allocation for the current year plus carryover from the previous fiscal year. Numbers may not total due to rounding.

Table 13 shows FWS's allocations in 2005, 2006 and 2007 to its seven regions, the California-Nevada Operations office, and headquarters.[2] In 2007, about 61 percent of FWS's total fuel reduction funding was allocated to WUI areas, while about 39 percent was allocated to non-WUI areas. In 2005, 2006, and 2007, the Southeast region received the most funding, followed by the Great Lakes-Big Rivers region.

Table 13. FWS Allocations to Regions and Headquarters, Fiscal Years 2005, 2006, and 2007

Region	Total allocation	Percentage of FWS's	Total WUI	Total non-WUI
Southeast				
2005	$7,005,484	24.7	$3,715,115	$3,290,369
2006	7,966,857	23.8	4,780,616	3,186,241
2007	7,543,624	24.2	4,481,330	3,062,294
Great Lakes-Big Rivers				
2005	4,867,717	17.1	2,111,460	2,756,257
2006	5,438,168	16.2	2,658,862	2,779,306
2007	5,336,376	17.1	2,604,636	2,731,740
Mountain-Prairie				
2005	3,376,546	11.9	1,281,699	2,094,847
2006	3,776,901	11.3	1,658,254	2,118,647
2007	3,690,940	11.8	1,566,284	2,124,656
Southwest				
2005	3,363,824	11.9	2,139,589	1,224,235
2006	3,903,921	11.7	2,499,122	1,404,799
2007	3,721,205	11.9	2,434,533	1,286,672
Pacific				
2005	2,556,707	9.0	1,504,681	1,052,026
2006	2,853,522	8.5	1,864,739	988,783
2007	2,566,156	8.2	1,736,153	830,003
Northeast				
2005	2,201,297	7.8	1,751,683	449,614
2006	2,597,811	7.8	2,106,088	491,723
2007	2,416,798	7.7	1,977,669	439,129
California-Nevada[a]				
2005	1,712,138	6.0	986,919	725,219
2006	2,542,027	7.6	1,960,241	581,786
2007	2,254,492	7.2	1,460,866	793,626
Alaska				
2005	1,000,439	3.5	815,607	184,832
2006	1,271,882	3.8	1,050,646	221,236
2007	1,224,552	3.9	1,074,688	149,864
Subtotal, regions				
2005	**26,084,152**	**91.9**	**14,306,753**	**11,777,399**
2006	**30,351,089**	**90.6**	**18,578,568**	**11,772,521**
2007	**28,754,143**	**92.1**	**17,336,159**	**11,417,984**
Washington Office				
2005	2,302,509	8.1	1,571,629	730,880
2006	3,153,074	9.4	2,037,897	1,115,177
2007	2,450,015	7.9	1,723,281	726,734
Total				
2005	**28,386,661**	**100**	**15,878,382**	**12,508,279**
2006	**33,504,163**	**100**	**20,616,465**	**12,887,698**
2007	**31,204,158**	**100**	**19,059,440**	**12,144,718**

Source: GAO analysis of FWS data.

Notes: Total allocations include the allocation for the current year plus carryover from the previous fiscal year. Numbers may not total due to rounding.

[a]While FWS has seven regions, it has an eighth office—the California-Nevada Operations office—that, although officially part of the Pacific region, manages its own fuel reduction program.

APPENDIX III: SUMMARY OF FUEL TREATMENT ACCOMPLISHMENTS FOR THE FOREST SERVICE AND INTERIOR, FISCAL YEARS 2005 AND 2006

The tables in this appendix summarize the fuel reduction accomplishments of the Forest Service and the four Interior agencies we reviewed—Bureau of Land Management (BLM), Bureau of Indian Affairs (BIA), National Park Service (NPS), and Fish and Wildlife Service (FWS)— for 2005 and 2006,[1] the most recent years for which complete data were available. National and regional office data are presented for each agency.[2]

Table 14 provides nationwide information for each of the five agencies, including total acres treated; acres treated in the wildland-urban interface (WUI) and in non-WUI areas; and acres treated with prescribed fire, mechanical methods, and other treatment methods such as herbicides and grazing. The Forest Service treated more acres than the four Interior agencies combined—almost 1.7 milli on acres in 2005 and more than 1.5 million acres in 2006. Within Interior, BLM and FWS treated the most acres, with BLM treating more than 500,000 acres in 2005 and almost 430,000 acres in 2006, and FWS treating almost 420,000 acres in 2005 and more than 370,000 acres in 2006. In each year, about 60 percent of the total acres treated by the five agencies were in the WUI, and about 40 percent of total acres treated were outside of the WUI. The majority of acres were treated with prescribed fire—almost 75 percent in 2005 and almost 65 percent in 2006.

Table 14. Summary of Fiscal Years 2005 and 2006 Fuel Reduction Accomplishments for Interior and Forest Service

Agency	Treated acres	WUI acres	Non-WUI acres	Acres treated using pre-scribed fire	Acres treated using mechanical means	Acres treated using other means
BLM						
2005	506,168	253,001	253,167	194,553	211,852	99,763
2006	427,912	230,932	196,980	107,443	206,123	114,346
BIA						
2005	193,617	71,983	121,634	96,881	94,168	2,568
2006	187,653	89,961	97,692	78,304	106,204	3,145
NPS						
2005	153,972	58,873	95,099	139,455	13,036	1,481
2006	116,635	38,558	78,077	102,765	11,532	2,338
FWS						
2005	415,646	158,711	256,935	389,686	21,734	4,226
2006	373,933	173,113	200,820	333,038	32,118	8,777
Interior subtotal						
2005	**1,269,403**	**542,568**	**726,835**	**820,575**	**340,790**	**108,0**
2006	**1,106,133**	**532,564**	**573,569**	**621,550**	**355,977**	**128**

Table 14. (Continued)

Forest Service						
2005	1,672,909	1,198,663	474,246	1,366,988	303,002	2,919
2006	1,503,475	1,090,721	412,754	1,061,277	433,077	9,121
Total Forest Service and Interior						
2005	**2,942,312**	**1,741,231**	**1,201,081**	**2,187,563**	**643,792**	**110,957**
2006	**2,609,608**	**1,623,285**	**986,323**	**1,682,827**	**789,054**	**137,727**

Source: GAO analysis of Interior and Forest Service data.

Table 15 summarizes 2005 and 2006 fuel treatment information for the Forest Service regions. In both years, the Southern region treated substantially more acres than the other regions, treating more than half of the Forest Service's than twice as many acres in the WUI as in non-WUI areas, and treated most acres with prescribed fire.

Table 15. Summary of Fiscal Years 2005 and 2006 Fuel Reduction Accomplishments for Forest Service Regions

Region	Treated acres	WUI acres	Non-WUI acres	Acres treated using prescribed fire	Acres treated using mechanical means	Acres treated using other means[a]
Pacific Southwest						
2005	100,540	59,194	41,346	41,565	58,785	190
2006	95,729	62,229	33,500	36,779	50,423	8,527
Southwestern						
2005	164,506	69,929	94,577	118,326	46,180	0
2006	180,616	84,973	95,643	134,289	46,327	0
Southern						
2005	976,176	803,654	172,522	969,528	6,616	32
2006	776,145	674,189	101,956	625,605	150,540	0
Pacific Northwest						
2005	139,470	79,975	59,495	65,955	73,515	0
2006	133,528	60,904	72,624	73,068	60,432	28
Rocky Mountain						
2005	93,969	65,862	28,107	51,353	39,919	2,697
2006	102,953	77,650	25,303	49,313	53,394	246
Intermountain						
2005	74,676	34,163	40,513	47,077	27,599	0
2006	87,957	33,995	53,962	58,075	29,562	320
Northern						
2005	70,594	43,824	26,770	39,726	30,868	0
2006	68,639	46,892	21,747	43,480	25,159	0
Eastern						
2005	51,472	40,556	10,916	33,089	18,383	0
2006	57,221	49,202	8,019	40,242	16,979	0
Alaska						
2005	1,506	1,506	0	369	1,137	0
2006	687	687	0	426	261	0
Total						
2005	1,672,909	1,198,663	474,246	1,366,988	303,002	2,919
2006	1,503,475	1,090,721	412,754	1,061,277	433,077	9,121

Source: GAO analysis of Forest Service data.

[a]"Other" category includes treatments such as herbicides and grazing.

Table 16 summarizes 2005 and 2006 fuel treatment information for the BLM state offices. In both years, the Oregon/Washington and Idaho state offices treated the most acres, followed by the New Mexico state office. BLM treated about the same number of acres in the WUI and in non-WUI areas in 2005, and treated about 34,000 more acres in the WUI than in non-WUI areas in 2006. Unlike the Forest Service, BLM treated more acres mechanically than with prescribed fire, and also treated a substantial number of acres using other treatment methods, such as herbicides or grazing.

Table 16. Summary of Fiscal Years 2005 and 2006 Fuel Reduction Accomplishments for BLM State Offices

State office	Treated acres	WUI acres	Non-WUI acres	Acres treated using prescribed fire	Acres treated using mechanical means	Acres treated using other means[a]
Oregon/ Washington						
2005	108,909	71,218	37,691	47,273	61,636	0
2006	92,918	69,521	23,397	30,138	62,762	18
Idaho						
2005	112,254	54,460	57,794	13,321	46,231	52,702
2006	113,778	69,648	44,130	12,307	47,614	53,857
Utah						
2005	40,706	26,616	14,090	6,140	33,966	600
2006	40,535	28,906	11,629	3,900	36,601	34
California						
2005	24,191	21,439	2,752	2,342	19,349	2,500
2006	19,389	16,231	3,158	4,187	11,422	3,780
Nevada						
2005	28,427	15,190	13,237	10,391	16,272	1,764
2006	35,465	9,655	25,810	15,242	15,014	5,209
Colorado						
2005	20,417	13,616	6,801	6,950	13,012	455
2006	17,870	10,132	7,738	4,906	12,774	190
New Mexico						
2005	48,107	3,397	44,710	21,297	7,704	19,106
2006	53,329	4,390	48,939	11,682	5,643	36,004
Montana						
2005	10,867	6,124	4,743	4,577	5,340	950
2006	12,446	7,530	4,916	5,910	6,426	110
Arizona						
2005	35,424	17,078	18,346	17,297	4,391	13,736
2006	19,557	7,424	12,133	5,845	3,625	10,087
Wyoming						
2005	30,839	1,976	28,863	19,885	3,004	7,950
2006	18,662	4,507	14,155	9,816	3,789	5,057
Alaska						
2005	45,707	21,847	23,860	45,080	627	0
2006	3,963	2,988	975	3,510	453	0
Eastern states						
2005	320	40	280	0	320	0
2006	0	0	0	0	0	0
Total						
2005	**506,168**	**253,001**	**253,167**	**194,553**	**211,852**	**99,763**
2006	**427,912**	**230,932**	**196,980**	**107,443**	**206,123**	**114,346**

Source: GAO analysis of Interior data. [a]"Other" category includes treatments such as herbicides and grazing.

Table 17 summarizes 2005 and 2006 fuel treatment information for the BIA regions. The Northwest and Western regions treated the most acres in 2005, with each region treating more than 38,000 acres. In 2006, the Northwest and Southwest regions treated the most acres, with each region treating more than 45,000 acres. The agency treated more acres in non-WUI areas than in WUI areas in 2005 and 2006.

Table 17. Summary of Fiscal Years 2005 and 2006 Fuel Reduction Accomplishments for BIA Regions

Region	Treated acres	WUI acres	Non-WUI acres	Acres treated using prescribed fire	Acres treated using mechanical means	Acres treated using other means[a]
Northwest						
2005	38,284	17,297	20,987	15,589	21,488	1,207
2006	48,733	25,354	23,379	14,687	32,590	1,456
Southwest						
2005	28,212	11,839	16,373	2,678	24,433	1,101
2006	45,132	20,558	24,574	8,096	36,049	987
Western						
2005	38,753	16,210	22,543	13,813	24,880	60
2006	22,167	10,360	11,807	7,892	14,275	0
Pacific						
2005	2,584	1,817	767	180	2,404	0
2006	4,431	3,179	1,252	331	4,100	0
Great Plains						
2005	14,386	6,986	7,400	7,595	6,591	200
2006	13,234	4,828	8,406	6,217	6,508	509
Alaska						
2005	1,253	1,253	0	167	1,086	0
2006	2,222	1,497	725	563	1,659	0
Midwest						
2005	21,356	6,478	14,878	17,792	3,564	0
2006	18,559	16,401	2,158	15,585	2,974	0
Rocky Mountain						
2005	11,347	2,856	8,491	6,616	4,731	0
2006	7,400	3,634	3,766	4,177	3,223	0
Navajo						
2005	14,274	956	13,318	13,318	956	0
2006	11,065	470	10,595	10,595	470	0
Southern Plains						
2005	12,322	434	11,888	8,401	3,921	0
2006	8,796	672	8,124	4,770	3,833	193
Eastern						
2005	7,788	5,616	2,172	7,718	70	0
2006	4,607	2,547	2,060	4,099	508	0
Eastern Oklahoma						
2005	3,058	241	2,817	3,014	44	0
2006	1,307	461	846	1,292	15	0
Total						
2005	**193,617**	**71,983**	**121,634**	**96,881**	**94,168**	**2,568**
2006	**187,653**	**89,961**	**97,692**	**78,304**	**106,204**	**3,145**

Source: GAO analysis of Interior data.

[a]"Other" category includes treatments such as herbicides and grazing.

Table 18 summarizes 2005 and 2006 fuel treatment information for the NPS regions. In 2005 and 2006, the Southeast region treated the most acres, followed by the Intermountain region. NPS treated more acres in non-WUI areas than the WUI, and treated the vast majority of acres (more than 90 percent in 2005 and about 88 percent in 2006) using prescribed fire.

Table 18. Summary of Fiscal Years 2005 and 2006 Fuel Reduction Accomplishments for NPS Regions

Region	Treated acres	WUI acres	Non-WUI acres	Acres treated using prescribed fire	Acres treated using Mechanical means	Acres treated using other means[a]
Pacific West						
2005	25,949	9,432	16,517	18,922	5,681	1,346
2006	22,433	9,042	13,391	16,220	4,101	2,112
Intermountain						
2005	43,823	28,585	15,238	38,874	4,844	105
2006	25,350	14,447	10,903	19,397	5,727	226
Southeast						
2005	63,602	17,963	45,639	62,491	1,081	30
2006	45,471	9,413	36,058	44,641	830	0
Midwest						
2005	20,082	2,433	17,649	18,971	1,111	0
2006	22,872	5,432	17,440	22,150	722	0
Northeast						
2005	453	417	36	188	265	0
2006	486	224	262	348	138	0
Alaska						
2005	29	9	20	0	29	0
2006	23	0	23	9	14	0
National Capital						
2005	34	34	0	9	25	0
2006	0	0	0	0	0	0
Total						
2005	**153,972**	**58,873**	**95,099**	**139,455**	**13,036**	**1,481**
2006	**116,635**	**38,558**	**78,077**	**102,765**	**11,532**	**2,338**

Source: GAO analysis of Interior data.

[a]"Other" category includes treatments such as herbicides and grazing.

Table 19 summarizes 2005 and 2006 fuel treatment information for the FWS regions. In both years, the Southeast region treated substantially more acres than the other regions—about 35 percent of total acres treated in 2005 and about 30 percent in 2006—followed by the Southwest and Great Lakes-Big Rivers regions. Like NPS, FWS treated most acres outside of the WUI, and treated the vast majority of acres (about 94 percent in 2005 and about 89 percent in 2006) using prescribed fire.

Table 19. Summary of Fiscal Years 2005 and 2006 Fuel Reduction Accomplishments for FWS Regions

Region	Treated acres	WUI acres	Non-WUI acres	Acres treated using prescribed fire	Acres treated using Mechanical means	Acres treated using other means[a]
Southeast						
2005	144,902	83,218	61,684	141,616	3,202	84
2006	114,212	75,024	39,188	106,864	7,348	0
Great Lakes-Big Rivers						
2005	73,550	27,719	45,831	70,880	2,204	466
2006	70,756	27,499	43,257	68,854	1,548	354
Mountain-Prairie						
2005	42,252	7,994	34,258	42,032	220	0
2006	39,095	10,100	28,995	38,862	233	0
Southwest						
2005	76,495	21,786	54,709	71,820	4,424	251
2006	56,607	17,036	39,571	54,792	1,800	15
Pacific						
2005	13,865	6,039	7,826	7,703	6,063	99
2006	23,996	15,704	8,292	11,225	11,683	1,088
Northeast						
2005	18,596	6,609	11,987	13,166	2,104	3,326
2006	16,515	8,791	7,724	13,007	1,794	1,714
California-Nevada[b]						
2005	45,216	4,595	40,621	42,176	3,040	0
2006	43,023	18,864	24,159	29,771	7,646	5,606
Alaska						
2005	770	751	19	293	477	0
2006	9,729	95	9,634	9,663	66	0
Total						
2005	**415,646**	**158,711**	**256,935**	**389,686**	**21,734**	**4,226**
2006	**373,933**	**173,113**	**200,820**	**333,038**	**32,118**	**8,777**

Source: GAO analysis of Interior data.

[a]"Other" category includes treatments such as herbicides and grazing.

[b]While FWS has only seven regions, it has an eighth office—the California-Nevada Operations office—that, although technically part of the Pacifi c region, manages its own fuel reduction program.

REFERENCES

[1] The National Fire Plan comprises multiple documents, including (1) a September 2000 report from the Secretaries of Agriculture and of the Interior to the President in response to the wildland fires of 2000, (2) congressional direction accompanying substantial new appropriations in fiscal year 2001, and (3) several strategies to implement all or parts of the plan. For a description of these documents and their contents, goals, and relationships to one another, see *Severe Wildland Fires: Leadership and Accountability Needed to Reduce Risks to Communities and Resources*, GAO-02-259 (Washington, D.C.: Jan. 31, 2002).

[2] Pub. L. No. 108-148 (2003).

[3] HFRA defines "federal land" to include land administered by the Forest Service and BLM. Consequently, HFRA fuel reduction project authorities are available only to the Forest Service and BLM, and its fuel reduction project requirements apply only to these agencies as well. In some cases, BIA, FWS, and NPS have chosen to comply with some of the requirements.

[4] Years cited in this report refer to fiscal years except where otherwise specified.

[5] GAO, *Western National Forests: A Cohesive Strategy Is Needed to Address Catastrophic Wildfire Threats*, GAO/RCED-99-65 (Washington, D.C.: Apr. 2, 1999); *Reducing Wildfire Threats: Funds Should Be Targeted to the Highest Risk Areas,* GAO/T-RCED-00-296 (Washington, D.C.: Sept. 13, 2000); GAO-02-259; *Wildland Fire Management: Additional Actions Required to Better Identify and Prioritize Lands Needing Fuels Reduction*, GAO-03-805 (Washington, D.C.: Aug. 15, 2003).

[6] GAO, *Wildland Fires: Forest Service and BLM Need Better Information and a Systematic Approach for Assessing the Risks of Environmental Effects*, GAO-04-705 (Washington, D.C.: June 24, 2004); *Wildland Fire Management: Important Progress Has Been Made, but Challenges Remain to Completing a Cohesive Strategy*, GAO-05-147 (Washington, D.C.: Jan. 14, 2005); *Wildland Fire Management: Update on Federal Agency Efforts to Develop a Cohesive Strategy to Address Wildland Fire Threats*, GAO-06-671R (Washington, D.C.: May 1, 2006).

[7] The Western Governors' Association independent, nonpartisan organization of governors representing 19 western states. The governors use the association to develop and advocate policies that reflect regional interests.

[8] The agencies treated an additional 1.8 million acres from 2004 to August 2007 through other land management activities.

[9] The agencies also conduct some treatments using other methods, such as applying herbicides and allowing animals to graze on the land.

[10] Interagency policy directs land managers to select firefighting strategies in accordance with local federal units' nt plans. If a plan has not been developedland and fire and approved, the policy directs land managers to suppress the fire. Thus, under the policy, the areas where wildland fire use is allowed must be defined in a fire management plan, along with prescribed weather and other conditions. The fires are monitored, and if weather conditions change in a way that would potentially allow the fires to escape from the designated areas, the fires are suppressed.

[11] This requirement applies only to projects conducted using HFRA authorities. Agency officials told us they do not have reliable data on the portion of their fuel reduction projects that used HFRA authorities.

[12] Our review was limited to fuel reduction work activities using federal funds appropriated specifically for this purpose. As a result, fuel reduction work funded by other agency programs or outside organizations is beyond the scope of this review.

[13] GAO/RCED-99-65.

[14] GAO, *Wildland Fire Management: Reducing the Threat of Wildland Fires Requires Sustained and Coordinated Effort*, GAO-02-843T (Washington, D.C.: June 13, 2002); GAO-05-147; GAO-06-671R; *Wildland Fire Management: Lack of a Cohesive Strategy Hinders Agencies'* , GAO-07-427T (Washington, D.C.: Jan. 30, 2007). *Cost-Containment Efforts*

[15] GAO/T-RCED-00-296; GAO-02-259.

[16] GAO-02-259; GAO-03-805.

[17] This approach, outlined by the National Academy of Public Administration, uses risk as a specific term referring to the probability of an event, as well as an umbrella term that encompasses all three of these elements. See National Academy of Public Administration, *Managing Wildland Fire: Enhancing Capacity to Implement the Federal Interagency Policy* (Washington, D.C.: December 2001).

[18] For more information on the hazard-risk-value framework, see GAO-04-705.

[19] GAO, *The National Fire Plan: Federal Agencies Are Not Organized to Effectively and Efficiently Implement the Plan*, GAO-01-1022T (Washington, D. C.: July 31, 2001).

[20] GAO-02-259.

[21] GAO-02-843T.

[22] The Forest Service's iorities for reducing hazardous fuels andsystem for setting allocating resources excluded the Alaska region (and did not give it a priority score) because the Alaska region's program for less than 1 percent of the agency's fuel reduction funds.

[23] See GAO-03-805; U.S. Department of Agriculture, Office of Inspector General, *Audit Report: Implementation of the Healthy Forests Initiative*, 08601-6-AT (Washington, D.C.: September 2006); Office of Management and Budget, *Program Assessment Rating Tool: Review of U.S. Department of Agriculture'sManagement Program* (Washington, D.C.: 2006); U.S. House of Representatives, Committee on Appropriations, *Department of the Interior, Environment, and Related Agencies Appropriations Bill of 2007, House Report 109-465*, 109th Cong., 2nd Sess. (Washington, D.C.: May 15, 2006); U.S. Senate, Committee on Appropriations, *Department of the Interior, Environment, and Related Agencies Appropriation Bill of 2007, Senate Report 109-275*, 109th Cong., 2nd Sess. (Washington, D.C.: June 29, 2006).

[24] The process the officials used to weight the factors is called the analytical hierarchy process, which is a systematic process often used in private industry to make complex decisions involving multiple criteria, such as investment decisions. We did not assess the appropriateness of the factors selected or the weights assigned, nor did we evaluate the model's ors to determine priority scores.accuracy in applying these fact

[25] GAO has previously noted the importance of the integrity, credibility, and quality of data that underlie budget decisions and, thus, the value in implementing changes gradually when the quality of such data is in question. See, for example, GAO, *Performance Budgeting: Opportunities and Challenges*, GAO-02-1106T (Washington, D.C.: Sept. 19, 2002).

[26] To ensure that the elements without data had no effect on priority scores, all of the regions were assigned the same score for each of these elements.

[27] The funds Interior allocated using the model represented 5 percent of project funds—that is, funds expected to be spent on individual projects—rather than 5 percent of the total allocation, which would include program management expenses such as salaries, facility costs, and so forth. NPS did not receive any of the 5 percent of 2007 funds that were allocated using the new model because Interior allocated the funds late in the fiscal year and NPS had already met its 2007 acreage targets.

[28] Weights are not included in the list because Interior had not yet finalized them at the time of our review.

[29] While agency officials told us the new model will be coordinated with FPA, they did not provide details on how this coordination will occur.

[30] While there are 12 BLM state offices—11 in the West and 1 in the East—the vast majority of BLM-managed land is in the West, and the Eastern States office receives only about 0.1 percent of BLM's total fuel reduction one field unit. As a result, the Eastern States office is not included in our description of BLM state office allocation processes.

[31] Regions that accomplished 90 percent or more of the previous year's acreage target could request up to 120 percent of the prior year's accomplished less than 90 percent could request only up to 105 percent of the previous year's amount.

[32] U.S. Department of the Interior and USDA Forest Service, "Protecting People and Natural Resources: A Cohesive Fuels Treatment Strategy," February 2006. Note that, although the document is referred to as a cohesive strategy, previous GAO reports concluded that it does not contain all the elements GAO called for in its earlier recommendations for such a strategy. See, for example, GAO-06-671R.

[33] The agencies expect that nationally consistent data will be available through LANDFIRE, a geospatial data and modeling system currently being implemented. LANDFIRE data are complete for some of the country, with data for the remainder of the country expected to be completed by 2009.

[34] In 2001, a *Federal Register* notice was published with a list of wildland-urban interface communities identified by states as being "in the vicinity of federal lands" and "at high risk from wildfire." However, the states and tribes used inconsistent approaches to identify these communities at risk. To standardize these approaches, the National Association of State Foresters was tasked, in the 10-Year Implementation Plan, with developing a definition for community at risk, and a process for states and tribes to follow to identify and prioritize the communities. Accordingly, in 2003, the National Association of State Foresters finalized its guidance, defining community as "a group of people living in the same locality and under the same government," and specifying that a community was to be considered at risk from wildland fire if it was located within the wildland-urban interface as defined in the 2001 *Federal Register*, which stated that, "the urban-wildland interface community exists where humans and their development meet or intermix with wildland fuel." We conducted our analysis using census data on the average population per square mile across areas defined by ZIP codes. However, especially in larger ZIP codes, there may be smaller pockets where the population density is higher or lower than the average used in our analysis.

[35] However, in a fire behavior assessment of the June 2007 Angora Fire in California, Forest Service officials stated that a large number of houses ignited because of embers from other burning houses, rather than from wildland fuel—suggesting that even homes that are not immediately adjacent to federal lands could be at risk from wildland fire.

Appendix I

[1] Years cited in this appendix refer to fiscal years except where otherwise specified.

[2] BIA, FWS, and NPS have regional offices, while BLM has state offices. For the purposes of this appendix, we refer to all of these as regional offices when we discuss the Interior agencies collectively.

[3] Results from nonprobability samples cannot be used to make inferences about a population, because in a nonprobability sample, some elements of the population being studied have no chance or an unknown chance of being selected as part of the sample.

[4] We conducted our analysis using U.S. Census data on the average population per square mile across areas defined by ZIP codes. However, especially in larger ZIP codes, there may be small pockets where the population density is higher or lower than the average used in our analysis.

[5] In the 2001 *Federal Register*, the agencies provide three categories of wildland-urban interface communities. The first is "interface community," which exists where structures directly abut wildland fuel; an alternative definition of the interface community specifies a population density of 250 or more people per square mile. The second is "intermix community," which exists where structures are scattered throughout a wildland area; an alternative definition of intermix community specifies a population density of between 28-250 people per square mile. The third is "occluded community," where structures, often within a city, abut an island of wildland fuel (e.g., park or open space).

Appendix II

[1] Years cited in this appendix refer to fiscal years except where otherwise specified. BIA, FWS, and NPS have regional offices, while BLM has state offices. For the purposes of this appendix, we refer to all of these as regional offices when we discuss the Interior agencies collectively.

[2] While FWS has seven regions, it has an eighth office—the California-Nevada Operations Office—that, although officially part of the Pa cific region, manages its own fuel reduction program.

Appendix III

[1] Years cited in this appendix refer to fiscal years except where otherwise specified.

[2] BIA, FWS, and NPS have regional offices, while BLM has state offices. For the purposes of this appendix, we refer to all of these as regional offices when we discuss the Interior agencies collectively.

Chapter 2

FOREST SERVICE: BETTER PLANNING, GUIDANCE AND DATA ARE NEEDED TO IMPROVE MANAGEMENT OF THE COMPETITIVE SOURCING PROGRAM [*]

WHY GAO DID THIS STUDY

Competitive sourcing is aimed at promoting competition between federal employees and the private sector as a way to improve government operations. Key work activities—those that are either inherently governmental or core to the agency's mission—are generally exempt from competitions. In fiscal year 2004, Congress began placing spending limitations on the Forest Service's competitive sourcing program because of concerns about how the program was managed. Also, like other agencies, the Forest Service must report annually to Congress on the savings achieved from any competitions it conducted.

GAO was asked to determine the extent to which the Forest Service has (1) plans and guidance to help implement its competitive sourcing program effectively and (2) sufficient cost data to ensure that it complied with its spending limitations and accurately reported its savings to Congress for fiscal years 2004 through 2006. To answer these objectives, GAO examined the agency's strategic plan, guidance, and available cost data for competitive sourcing and interviewed key agency officials.

ABBREVIATIONS

CSO	competitive sourcing officer
CSPO	Competitive Sourcing Program Office
FFIS	Federal Financial Information System
FTE	full-time-equivalent staff
ICS	Incident Command System
IT	information technology

[*] This Chapter is an excerpted indexed edition of GAO Report GAO-08-195, Dated January 2008

MEO	Most Efficient Organization
NWCG	National Wildfire Coordinating Group
OCFO	Office of the Chief Financial Officer
OMB	Office of Management and Budget
PWS	performance work statement
USDA	U.S. Department of Agriculture

January 22, 2008
Congressional Requesters

The 2001 *President's Management Agenda* includes a governmentwide initiative—known as competitive sourcing—that aims to improve government efficiency and reduce the costs of government programs by promoting competition between federal employees and private sector organizations.[1] The competitions determine who should perform commercial activities—activities performed by the government that are also regularly performed in the commercial marketplace, such as information technology (IT), maintenance and property management, and logistics. These competitions compare the public and private sector costs of performing an activity and may result in outsourcing federal jobs to the private sector. In its competitive sourcing strategic plan, known as the Green Plan, the Forest Service, an agency within the U.S. Department of Agriculture (USDA), sets forth the agency's goal to conduct 14 precompetition studies in 5 years to evaluate the feasibility of subjecting nearly two thirds of its total workforce to competitive sourcing competitions.

The Forest Service manages about 193 million acres of land—about 30 percent of all federal lands—and has a workforce of about 37,000 full-time-equivalent staff (FTE). The employees are geographically dispersed throughout the nation. In addition to its headquarters in Washington, D.C., the Forest Service has 9 regional offices, 155 forest offices, and about 600 district offices.

The Forest Service's competitive sourcing program is subject to the requirements of the Office of Management and Budget's (OMB) Circular No. A-76 (the circular) and the Federal Activities Inventory Reform Act of 1998 (the FAIR Act).[2] Generally, the circular sets forth the competitive process an agency must follow to determine whether federal employees or private sector organizations should perform a commercial activity. The competitive process can be divided into three stages: (1) a precompetition planning stage that, among other things, determines the scope of the competition; (2) a competition stage that begins with a public announcement of a competition and ends with the selection of the competition's winner; and (3) a postcompetition accountability stage, which involves such activities as monitoring and reporting on the winner's performance. The circular also directs agencies to create a staffing plan—which creates an organization of agency employees that is known as the "Most Efficient Organization" (MEO)—to be used to represent the agency in the competition. After allowing for a transition period, the winner of the competition, whether the MEO or a private sector organization, is subject to the provisions of the postcompetition accountability stage.

The FAIR Act, as implemented by the circular, requires federal agencies to compile and submit to OMB for approval an annual inventory of the activities their employees perform. The inventory must include the number of FTEs that perform each activity and designate whether the activity is inherently governmental or commercial. Inherently governmental activities—such as setting policy or spending funds—are critical to the role and function of government and, therefore, are exempt from competition with the private sector. In contrast, commercial activities—those that are not inherently governmental and could be performed by the private sector—are subject to competitive sourcing competitions, unless the agency exempts the activity. The agency may still exempt a commercial activity if, for example, it deems the activity to be core to the agency's mission (referred to as core-commercial).

In addition, the Consolidated Appropriations Act, 2004, requires executive agencies to report annually to Congress on competitive sourcing activities.[3] Among other things, the act requires agencies to report actual savings derived from implementing a competitive sourcing competition. While the act does not provide direction on how to calculate actual savings, OMB does provide this direction through annual guidance.[4] The Forest Service, through USDA, reported to Congress that it saved over $38 million in fiscal years 2004 through 2006 as a result of competitive sourcing competitions.

Because of concerns about how the Forest Service was managing its competitive sourcing program, Congress limited the funds that the Forest Service could spend on competitive sourcing activities during fiscal years 2004 through 2007. In fiscal year 2004, the spending limitation was $5 million; in fiscal year 2005, it was $2 million; in fiscal year 2006, it was $3 million; and in fiscal year 2007, it was also $3 million.[5] The Forest Service initiated 171 competitions in fiscal years 2002 and 2003, and it has acknowledged that this first experience with the competitive sourcing initiative fell short of expectations. For example, a congressional study noted that nearly half of these competitions involved three or fewer FTEs, and that, as a result, Forest Service officials acknowledged that the competitions were unrealistic, adding that holding competitions involving few FTEs often made it impossible for private sector organizations to compete effectively.[6] Two of the most important lessons the Forest Service believes it has learned from this experience are that it should (1) centralize responsibilities for competitions to increase control and oversight and (2) plan competitions better by, for example, identifying the appropriate scope of the competition.

The Forest Service has taken steps to improve the management of its competitive sourcing program in both of these areas. In fiscal year 2003, it strengthened headquarters' oversight of competitive sourcing competitions. In fiscal year 2004, USDA issued guidance directing the Forest Service to perform studies to assess the feasibility of conducting a competition—known as feasibility studies—before committing to a competition. Although not required by the circular, feasibility studies are conducted during the precompetition planning stage and help determine whether to compete an activity. Also, some of the steps required by the circular during the precompetition planning stage are completed as part of the study. In fiscal years 2004 through 2006, the Forest Service completed three feasibility studies. They were for communications activities (completed June 2005), the Geospatial Service and Technology Center (completed April 2006),[7] and fleet management activities (completed May 2006). During this time frame, the Forest Service also completed three competitions, all begun in 2003 and completed in 2004. They were the IT competition (1,200 FTEs, completed July 2004); the fleet maintenance competition (57 FTEs, completed January 2004); and the road maintenance competition (66 FTEs, completed January 2004).[8]

As with the management of its competitive sourcing program in fiscal years 2002 and 2003, the Forest Service's most recent competitive sourcing activities have raised questions about whether the program is being effectively managed. In this context, you asked us to determine the extent to which the Forest Service has (1) plans and guidance to help implement its competitive sourcing program effectively and (2) sufficient cost data to ensure that it complied with its competitive sourcing statutory spending limitations and accurately reported its competitive sourcing savings to Congress for fiscal years 2004 through 2006.

To address these two objectives, we focused on the Forest Service's competitive sourcing activities between fiscal years 2004 through 2006. For the first objective, we examined the Forest Service's Green Plan and how the Forest Service scoped the three competitions it completed during this period. In doing so, we collected and reviewed pertinent OMB, USDA, and Forest Service guidance and documentation, including documentation related to the three completed competitions. In addition, we interviewed Forest Service officials familiar with the agency's competitive sourcing activities to better understand implementation of the program. For the second objective, we examined the appropriations acts for fiscal years 2004 through 2006 to determine which costs incurred by the Forest Service during that period were subject to the spending limitations. We obtained available Forest Service cost data for activities associated with its competitive sourcing program. We assessed the reliability of these data and found them unreliable as a measure of the costs and savings associated with the program. We conducted our work between October 2006 and January 2008 in accordance with generally accepted government auditing standards. See appendix I for a more detailed description of our objectives, scope, and methodology.

RESULTS IN BRIEF

The Forest Service lacks a realistic strategic plan and adequate guidance to help ensure that it can effectively implement its competitive sourcing program. Specifically, the Forest Service's Green Plan does not identify the personnel and funding resources that are likely to be available to implement the plan. These personnel and resource demands are likely to be substantial because the plan proposes to subject all commercial activities—affecting about two thirds of the agency's FTEs—to feasibility studies. This includes a single "catch-all" feasibility study covering 15,000 FTEs that was not included in a draft of the Green Plan because the activities associated with these FTEs did not satisfy the Forest Service's initial screening criteria for identifying good candidates for feasibility studies. The Forest Service added the additional "catch-all" feasibility study to its Green Plan because it was directed by USDA to schedule feasibility studies for all eligible FTEs without taking into account resource considerations. Furthermore, although the Forest Service identified inherently governmental and core-commercial activities in its annual FAIR Act inventories, Forest Service officials responsible for planning the three competitions completed in fiscal year 2004 did not find the inventory data useful for identifying and exempting those activities because, among other things, the officials believed that the classifications used in the FAIR Act inventories did not capture the actual work activities Forest Service employees carried out. Although the officials said they received little guidance on how to supplement the FAIR Act inventories or how to identify inherently governmental and core-commercial activities, they do not believe that these inappropriate activities were included in the competitions because of the relatively small size and clearly commercial nature of these competitions. However, without clear guidance, and in light of its plans to examine the activities of two thirds of its workforce, we believe that the agency is at risk of subjecting inherently governmental and core-commercial activities to future competitive sourcing competitions. In addition, the Forest

Service lacks a strategy on how to assess the cumulative effect that competing activities could have on its ability to fight wildland fires and respond to other emergencies. Forest Service employees play a prominent role in the nation's ability to fight wildland fires and respond to emergencies. Outsourcing substantial numbers of Forest Service jobs to the private sector could, over time, reduce this ability.

For fiscal years 2004 through 2006, we found that the Forest Service lacked sufficiently complete and reliable cost data to (1) demonstrate its compliance with appropriations acts' spending limitations on its competitive sourcing activities and (2) accurately report competitive sourcing savings to Congress. Regarding compliance with the statutory spending limitations, the Forest Service did not collect cost data on all activities related to competitive sourcing because it interpreted the legislation narrowly—believing that some costs were not subject to the spending limitations. For example, the Forest Service did not collect data on the cost of employee salaries related to feasibility studies—a key component of its competitive sourcing process. However, in our view, the appropriations acts' spending limitations, with one exception, apply to all costs attributable to the Forest Service's competitive sourcing program—including, for example, the costs of salaries related to feasibility studies. The one exception to the spending limitations is the cost of activities incurred to carry out the FAIR Act, such as preparing inventories, because this act requires the Forest Service to perform these activities even in the absence of other competitive sourcing activities. (See app. II for our full legal interpretation.) Moreover, the Forest Service failed to ensure the reliability of the cost data that it did collect to determine compliance with statutory spending limitations. As a result, the Forest Service cannot know if it exceeded the appropriations acts' spending limitations, even using its own narrow interpretation of which costs are subject to the spending limitations. Regarding savings reported to Congress, the Forest Service could not provide us with sufficient data or the methodology it used to calculate savings derived from competitions, and, as a result, we could not verify the accuracy of the reported savings. In addition, we found that the Forest Service did not consider certain substantial costs in its savings calculations, and thus Congress may not have an accurate measure of the savings produced by the Forest Service's competitive sourcing competitions. Although OMB provides guidance on how to calculate the savings, the guidance does not specify all of the costs that should be included in the calculations, thus providing the Forest Service with some discretion on which costs to include. Some of the costs the Forest Service did not include in the calculations substantially reduce or even exceed the savings reported to Congress. For example, regarding the IT infrastructure competition, the Forest Service did not include the $40 million that it cost to make the transition to the MEO. This amount is $5 million more than the $35 million in savings that the agency reported to Congress.

We are recommending that the Secretary of Agriculture direct the Chief of the Forest Service to improve how the agency manages its competitive sourcing program to help ensure that it has a realistic strategic plan, excludes inappropriate activities from competitive sourcing competitions, maintains its capacity to respond to wildland fires, and has the cost data necessary to comply with appropriations acts' spending limitations and to accurately report savings to Congress. In commenting on a draft of this report, the Forest Service generally agreed with our recommendations but had concerns about some of the specific findings and conclusions in the report. The agency's concerns and our responses to them are presented in appendix III.

BACKGROUND

The Forest Service's mission includes sustaining the nation's forests and grasslands, managing the productivity of those lands for the benefit of citizens, conserving open space, enhancing outdoor recreation opportunities, and conducting research and development. To help fulfill its mission, the Forest Service devotes considerable resources to suppressing wildfires. To coordinate the firefighting efforts of the Forest Service and other federal land management agencies, the interagency National Wildfire Coordinating Group (NWCG) was established.[9] This group adopted an interagency incident command system (ICS) and firefighting standards for responding to wildland fires. Federal employees in the land management agencies assume specific roles within the ICS, a command structure used at all levels of government and organized around five primary functional areas: command, operations, planning, logistics, and finance and administration.

There are about 80,000 federal employees and retirees who mobilize to assist state and local organizations to fight fires and respond to national emergencies as needed. These employees must receive standardized training and be certified in specific ICS duties (e.g., communications, aircraft management, and dispatch) before being available to respond to wildfires and other emergencies.

Approximately 30,000 permanent Forest Service employees are ICS-certified, with 10,000 to 12,000 of these employees holding fire-related positions. The remaining 18,000 to 20,000 employees are part of the "Forest Service militia." These are employees who are ICS-certified to fight fires but for whom the militia duty is a volunteer, collateral duty. These militia members will often leave their primary work duties and travel to the incident scene to provide assistance. A militia member may typically spend between 1 to 3 weeks each year fighting wildfires and responding to emergency situations. Militia teams that assisted the Federal Emergency Management Agency following Hurricane Katrina logged as many as 60 to 90 days of duty. As part of their militia duties, Forest Service employees often perform activities related to their regular work duties. For example, when responding to a wildfire, an employee whose regular duties involve IT support might provide this support at the scene of the fire. Employees may also perform militia duties that are unrelated to their regular work duties. For example, an IT employee who is ICS-certified in logistics might be responsible for providing food, supplies, and equipment at the scene of a fire.

Competitive Sourcing Policy and Process

The federal government has had a long-standing acquisition policy that, when permissible and cost-effective, agencies are to rely on the private sector to perform activities that are regularly performed in the commercial marketplace, such as IT, maintenance and property management, and logistics. This policy was laid out in OMB's 1966 Circular No. A-76, which was last revised in 2003.[10] The circular's stated goal is to obtain maximum value for taxpayers' dollars by taking advantage of competitive forces. This policy, described in the 2001 *President's Management Agenda* as competitive sourcing, is one of five governmentwide initiatives intended to improve the federal government's management and performance so that resources entrusted to the federal government are well managed and used wisely.[11] In addition, the circular provides agency management with a structured process to compare the public and private sector costs of performing an activity and to select the lowest cost provider through competition. This comparison may result in outsourcing federal jobs to the private sector.

The first step toward competitive sourcing is identifying work activities that are suitable for competition. The FAIR Act, as implemented by the circular, requires federal agencies to annually inventory all of the activities that federal employees perform—the FAIR Act inventory. For the inventory, activities are classified as inherently governmental or commercial. Specifically:

- *Inherently governmental activities* are those activities that are so intimately related to the public interest that they require performance by federal government employees.[12] The circular exempts these activities from competition.
- *Commercial activities* are those that are not inherently governmental and could be performed by the private sector. Commercial activities listed in the FAIR Act inventory are subject to the competition process detailed in the circular, unless they are placed in a subcategory of commercial activities that are exempt from competition.
- *Core-commercial activities* are those in a subcategory of commercial activities that are identified to be essential, or "core," to the agency's mission.[13] The circular allows agencies to exempt these activities from competition with sufficient written justification.[14]

To determine whether federal employees or private sector organizations should perform commercial activities, the circular establishes the following three-stage competitive process (referred to as an A-76 competition):

- *A precompetition planning stage* that, among other things, determines the scope of the competition. In this stage, the agency determines the commercial activities to be competed and the precompetition cost to perform that activity. It also appoints competition officials who will be in charge of developing, for example, the performance work statement (PWS), which specifies the work to be performed by the winning bidder. In addition, agency officials unaffiliated with the PWS create the MEO, which is generally a smaller, streamlined version of the government organization that is currently doing the work.
- *A competition stage* that begins with a public announcement of a competition and ends with the selection of the competition's winner. During this stage, an agency develops and issues a solicitation, receives offers, and follows a process to select the winning bidder.
- *A postcompetition accountability stage* that involves such activities as monitoring and reporting on the winner's performance, whether that winner is the MEO or a private sector organization.

The circular also directs agencies to designate a competitive sourcing official (CSO) with responsibility for implementing the circular who, with certain exceptions, may delegate those responsibilities to other officials in the agency. USDA designated its Chief Financial Officer as its CSO. He and his office, the Office of the Chief Financial Officer (OCFO), provide oversight to USDA and its agencies, including the Forest Service.

In 2002, the Forest Service established the Competitive Sourcing Program Office (CSPO) in its headquarters. The CSPO oversees the preparation of the FAIR Act inventory and provides written guidance to employees throughout the agency. In 2003, competitive sourcing responsibilities were further centralized in the CSPO to include oversight of A-76 competitions and postcompetition reporting. Supporting the CSPO are regional, national forest, and district office staffs.

Forest Service Green Plans and Feasibility Studies

OMB requires that agencies develop a strategic plan—known as a Green Plan—for implementing their competitive sourcing programs.[15] OMB's guidance on how to develop a Green Plan describes it as a long-range plan to ensure that competitive sourcing is a carefully and regularly considered option for improving the cost-effectiveness and quality of an agency's commercial activities. According to the guidance, agencies should include in the plan a description of how they are going to take timely and effective advantage of competition, a list of activities being announced for competition, an overview of their decision-making process, and a strategy to limit potential constraints on competition. The guidance also requires that agencies update their plans as organizational conditions change. To comply, the Forest Service has periodically submitted its Green Plan to USDA for incorporation into the department-level Green Plan, which is then presented to OMB for approval. Since December 2003, the Forest Service has submitted at least five versions of its Green Plan to USDA. The Forest Service's most recent OMB-approved Green Plan was issued in December 2005 and covers fiscal years 2005 through 2009.

In its Green Plan, the Forest Service proposes conducting a series of competitive sourcing feasibility studies before holding A-76 competitions. Feasibility studies enable an agency to first examine the practicality of subjecting activities to a competition before committing to one. OMB recognizes the value of this step and has recommended in its guidance that agencies conduct feasibility studies to streamline the competitive sourcing process. The OCFO issued guidance in May 2004 on conducting feasibility studies that outlined USDA agencies' responsibilities and specific procedures to be followed during a feasibility study.

The Forest Service typically assembles a team of six to eight employees to conduct a feasibility study. Each study usually involves one or more requests for data from field offices, which may require information from several hundred Forest Service employees. In addition to recommending whether to proceed with an A-76 competition, a Forest Service feasibility study can make other recommendations, such as to reorganize the way in which the Forest Service performs the activity being studied without engaging in an A-76 competition. When the feasibility study is completed, the Chief of the Forest Service reviews the study team's report and recommendations and decides on the best course of action, which may or may not be an A-76 competition.

In fiscal years 2004 through 2006, the Forest Service completed three A-76 competitions (see table 1).

Table 1. Forest Service's Completed Competitions, Fiscal Years 2004-2006

Activity	FTEs	Winner announcement date	Competition winner
Information technology infrastructure	1,200	July 2004	MEO
Fleet maintenance	57	January 2004	Private sector contractor
Road maintenance	66	January 2004	MEO

Source: USDA.

Competitive Sourcing Reporting Requirements and Spending Limitations

The Consolidated Appropriations Act, 2004, requires executive agencies, such as USDA, to report to Congress on their competitive sourcing activities for the prior fiscal year, including the total number of competitions announced and completed; the incremental costs directly attributable to conducting these competitions; and the total savings actually, or estimated to be, derived from such competitions. As an agency within USDA, the Forest Service must report this information to USDA for inclusion in USDA's report to Congress.

Congress also limited the Forest Service's funds available for "competitive sourcing studies and related activities" in each year's appropriations act for fiscal years 2004 through 2007, because of concerns about how the Forest Service had implemented its competitive sourcing initiative. Specifically, the spending limitations were $5 million (fiscal year 2004), $2 million (fiscal year 2005), $3 million (fiscal year 2006), and $3 million (fiscal year 2007).[16]

The Forest Service Does Not Have a Realistic Strategic Plan or Adequate Guidance to Effectively Implement Its Competitive Sourcing Program

The Forest Service lacks a realistic strategic plan and adequate guidance to help ensure that it can effectively and efficiently implement its competitive sourcing program. Specifically, the Forest Service's Green Plan proposes to subject all commercial activities to feasibility studies without identifying the personnel and funding resources that are likely to be available for the studies. Furthermore, Forest Service officials responsible for planning the three competitions completed in fiscal year 2004 told us that they did not find the FAIR Act inventory data useful for identifying and exempting inherently governmental and core-commercial activities, and that they received little guidance on how to supplement the inventory data or how to identify inherently governmental and core-commercial activities without using the inventory data. Nevertheless, Forest Service officials told us that the lack of guidance did not result in competing inherently governmental and core-commercial activities because of the small size and commercial nature of these competitions. However, without clear guidance, we believe that the agency risks subjecting inherently governmental and core-commercial activities to future A-76 competitions. In addition, the Forest Service does not have a strategy on how to assess the cumulative effect that competing activities could have on its ability to fight wildland fires and respond to other emergencies, even though outsourcing a large number of federal jobs to the private sector could reduce the availability of certified responders in the long term.

The Forest Service's Strategic Plan for Managing Its Competitive Sourcing Program Is Not Realistic

The Forest Service's December 2005 Green Plan for managing its competitive sourcing program is not realistic because it does not take into account the personnel and funding resources that are likely to be available to implement the plan, even though it proposes to subject all commercial activities—performed by approximately 24,500 FTEs—to feasibility studies during fiscal years 2005 through 2009. This is a significant increase over the activities it proposed to study in a draft of this plan issued 5 months earlier in July 2005. The July 2005 draft Green Plan scheduled 13 feasibility studies for activities associated with 6,180 FTEs for fiscal years 2005 through 2009. The Forest Service selected these activities by identifying good candidates for feasibility studies and then selecting a level of effort the officials believed could be managed over the next 5 years. For example, in selecting the studies, the officials said they considered the complexity of the studies, the additional workload that would result from conducting them, and the personnel resources required in selecting these activities. The Forest Service used the following nine criteria as the basis for identifying good candidates for feasibility studies:

- *Potential for savings.* An activity with greater potential for savings through more effective or efficient performance is considered a stronger candidate for a feasibility study.
- *Availability of private sector contractors.* An activity performed by a large number of commercial companies is considered a stronger candidate for a feasibility study.
- *Severability.* An activity that can be performed by an independent business unit is considered a stronger candidate for a feasibility study.
- *Preferred government performance.* An activity that management prefers to be performed by a government position is considered a weaker candidate for a feasibility study.
- *Location.* An activity that does not have to be performed locally is considered a stronger candidate for a feasibility study.
- *Fragmentation.* An activity that is the minor responsibility of a large number of positions, while difficult to compete, may benefit from restructuring and is considered a stronger candidate for a feasibility study.
- *Centrality of performance.* An activity that is typically performed at a central location is considered a stronger candidate for a feasibility study.
- *Potential for process improvement.* An activity with a greater potential for improvement through modernization, reorganization, or some other means is considered a stronger candidate for a feasibility study.
- *Impact on incident support.* An activity that supports emergency situations, such as firefighting, is considered a weaker candidate for a feasibility study.

However, after reviewing the Forest Service's proposed Green Plan, USDA directed the Forest Service to revise its plan to include all 24,512 commercial FTEs eligible for competition in either a feasibility study or an A-76 competition. In response, the Forest Service issued a revised Green Plan in December 2005, which OMB subsequently approved. The revised plan included a single "catch-all" feasibility study—labeled "All Other Commercial B Activities"—which had 15,000 FTEs associated with it,[17] nearly all of the commercial FTEs not already identified in the draft Green Plan. According to a Forest Service official, the additional feasibility study was added to comply with the USDA directive to include all commercial FTEs in the Green Plan, not because it included activities that might benefit from an A-76 competition. OCFO officials explained to us that USDA agencies have an option to perform feasibility studies to identify good candidates for more targeted feasibility studies, and that this was the purpose of the 15,000 FTE feasibility study. As of September 30, 2007, the Forest Service had not started this study.

As table 2 shows, the Forest Service had completed only five of the nine feasibility studies scheduled to be completed by September 30, 2007. The completed studies account for only 2,580 FTEs of the approximately 19,000 FTEs that were scheduled to be studied by this date.

Table 2. Forest Service's Schedule for Conducting Feasibility Studies and the Studies' Status as of September 30, 2007

Activities subject to feasibility study	FTEs	Proposed beginning date	Proposed completion date	Feasibility study begun	Feasibility study completed
Communication	750	April 2005	June 2005	Yes	Yes[a]
Aviation & Other Airborne Activities	500	October 2005	March 2006	Yes	Yes
Fleet Management	135	October 2005	March 2006	Yes	Yes
Geospatial Services & Tech Center	95	October 2005	March 2006	Yes	Yes
Computer Application Development	750	September 2006	March 2007	No	No
NEPA Information Collection & Analysis	1,100	September 2006	May 2007	Yes	Yes
Dispatch/Coordination System	300	September 2006	March 2007	Yes	No
All Other Commercial B Activities	15,000	October 2006	February 2007	No	No
Fire & Aviation Training	300	March 2007	September 2007	Yes	No
Fuels Management Program	500	September 2007	March 2008	No	
Technology or Service Centers	200	April 2008	September 2008		
Safety & Occupational Health	150	September 2008	March 2009		
Engineering Design & Development	900	September 2008	May 2009		
Fire Preparedness Program	500	November 2008	May 2009		
Total	**21,180**[b]				

Source: Forest Service.

[a]The Forest Service used a private consultant to perform the communication feasibility study. The study was completed but not accepted by the Forest Service because it did not meet OCFO requirements.

[b]While the revised Green Plan states that the agency plans to conduct feasibility studies on activities associated with 24,512 FTEs, the plan actually scheduled only studies for activities associated with 21,180 FTEs, with an additional 130 FTEs scheduled for an A-76 competition concerning communication activities. The remaining 3,202 FTEs are unaccounted for.

In our previous work,[18] we found that effective strategic plans take into account the resources required to implement the plan, such as human capital, technology, and information. While Forest Service officials could not provide us with any documents showing resources involved in conducting feasibility studies, the effort does not appear to be insignificant. According to OCFO guidance, 13 separate steps are involved in conducting feasibility studies, with many of the procedures involving several additional subtasks. Table 3 shows OCFO's guidance for conducting feasibility studies.

Table 3. USDA OCFO Guidance for Conducting Competitive Sourcing Feasibility Studies

Procedure	Description of activities involved
Business needs assessment	• Evaluate core public need for activity • Identify key factors that affect performance of activity • Align activity with agency's strategic goals and objectives
Assumptions and constraints	• List all assumptions regarding the work performed for the activity – For example, scope, timelines, workload requirements, necessary technology, customers, partners, stakeholders, funding, security, and outside support • List all constraints regarding the work performed for the activity – For example, time, budget, organizational, structure, and physical factors
Market research	• Analyze capabilities of commercial marketplace • Determine if alternative sources can satisfy the requirements of the activity • Create inventory of all commercial firms that produce, distribute, and support products and/or services similar to activity • Ensure that all market research complies with Federal Acquisition Regulation Part 10 requirements
Current "as-is" assessment	• Describe "as-is" scope of current operations – Discuss current workload, customers, partners, and stakeholders – Describe pros and cons of current operations – Describe workflow, required human capital resources – Provide overview of how internal functions work with one another – Describe potential effects on other USDA and non-USDA organizations should the activity be outsourced
Future "to-be" assessment	• Present a "to-be" scope of operations • Analyze expected workload, potential growth, future customers, and stakeholders • Develop "most-efficient organization" model to accomplish the activity
Performance gap analysis	• Define the gaps between the "as-is" and "to-be" scopes of operation • Prioritize performance gaps from most- to least-critical
Cost/Benefit analysis	o Compare cost and benefits of all options for performing the activity to determine the most cost- effective solution – Calculate cost estimate o For example, labor, employee benefits, materials, facilities, contractor expenses, and miscellaneous costs – Calculate benefits estimate o Estimate the overall public-value added (in dollars) of performing the activity – Perform savings analysis o Compare "bottom line" of each performance option

Table 3. (Continued)

Procedure	Description of activities involved
Civil rights impact assessment	• Identify and categorize the civil-rights impact of implementing competitive sourcing in accordance with USDA Office of Civil Rights policy guidance • Identify any potential risk of violation
Systems	• Provide an overview of all capital infrastructure systems and their relationships – Current systems used to support activity – System-to-system interfaces – Planned or in-progress system upgrades • Identify how the alternative options conform to the agency's IT architecture standards and guidelines
Acquisition strategy	• Develop resource acquisition plan for competitive sourcing option • Identify expected performance levels • Develop oversight mechanism to ensure that minimum performance standards are met
Project life-cycle schedule	• Identify major activities, milestones, and transition timelines from "as-is" to "to-be" system, including • all necessary tasks and subtasks related to the activity
Requirements	• List the items needed to complete the activity
Recommendations	• Recommend future course of action based on feasibility study results • If competitive sourcing is recommended, identify timelines set forth in the circular

Source: USDA OCFO, Guidance for Determining the Feasibility of Conducting Competitive Sourcing Competitions, Bulletin 2004-001 (May 11, 2004).

According to a senior Forest Service official, while the agency considered personnel requirements when selecting the original 13 activities for feasibility studies, it did not do so when it expanded the Green Plan to include all commercial FTEs. According to this official, there would be no practical way to conduct the 15,000-FTE study because it would include so many dissimilar activities. Consequently, FTEs would first need to be grouped into many separate activities, each requiring its own feasibility study, and as the number of studies increased, so too would the demands placed on Forest Service personnel. Several senior Forest Service officials with whom we spoke said that it is inconceivable that the schedule of feasibility studies in the OMB-approved Green Plan for fiscal years 2005 through 2009 could be met.

Just as it did not take into account personnel resources in its OMB-approved December 2005 Green Plan, the Forest Service also did not consider congressionally directed funding limitations. While the Forest Service's July draft Green Plan acknowledged the fiscal year 2005 $2 million statutory spending limitation on the Forest Service's competitive sourcing activities, the December Green Plan did not. In directing the Forest Service to include all commercial activities in the plan, USDA said to do so as if there were no funding limitations. Because Congress had placed limitations on the Forest Service's spending for competitive sourcing activities in the previous 2 fiscal years—2004 and 2005—factoring in the possibility of limited funds available in future years would have been appropriate.

The Forest Service Lacks Guidance to Ensure That Key Work Activities Are Excluded from A-76 Competitions

Since an agency's FAIR Act inventory designates all of an agency's FTEs as inherently governmental, commercial, or commercial but exempt from competition (e.g., core-commercial), the development of an accurate inventory becomes the foundation for determining which activities agencies select for competition. As we have previously reported,[19] other agencies have had difficultly in classifying positions when preparing their FAIR Act inventories, and the Forest Service is no exception. The Forest Service's difficulty is exemplified by significant fluctuations in the percentages of inherently governmental activities in its inventory data for fiscal years 2004 through 2006. In addition, there are differences between the Forest Service's initial classifications and those reported in the agency's OMB-approved Fair Act inventory. See table 4.

Table 4. Percentages of Inherently Governmental and Core-Commercial Activities Identified by the Forest Service Compared with the Percentages Reported in the OMB-Approved FAIR Act Inventories, Fiscal Years 2004-2006

Fiscal year	Inherently governmental activities		Core-commercial activities	
	Percentage identified by the Forest Service	Percentage reported in the Forest Service's OMB- approved FAIR Act inventory	Percentage Identified by the Forest Service	Percentage reported in the Forest Service's OMB- approved FAIR Act inventory
2004	21%	21%	14%	0%
2005	54	21	13	0
2006	7	7	13	13

Source: Forest Service.

According to a Forest Service official involved in preparing the agency's FAIR Act inventory, the fluctuations in the percentages of FTEs designated as inherently governmental—ranging from a high of over 50 percent to a low of 7 percent—were the result of changes from year to year in both the Forest Service's criteria for classifying activities, and the methodology it used to calculate the percentages of FTEs performing inherently governmental activities. In addition, the differences in the percentages of FTEs designated as core-commercial activities stemmed from disagreements between the Forest Service and OMB about what constituted a core-commercial activity. Specifically, in fiscal years 2004 and 2005, OMB did not approve the Forest Service's written justifications for core-commercial activities. Instead, it directed the Forest Service to reclassify all of the activities the agency had identified as core-commercial activities to commercial, thus making these activities eligible for competition.

The Forest Service's lack of consistency in its classification methodology, coupled with disagreement between the Forest Service and OMB regarding activity classifications, call into question the accuracy and usefulness of the Forest Service's FAIR Act inventory data for identifying inherently governmental and core-commercial activities when planning specific A-76 competitions. Forest Service officials told us that the FAIR Act inventory represents only a "rough snapshot" of the inherently governmental and core-commercial activities within the Forest Service, and that much additional work must be done to identify specific activities suitable for competition. They raised the following concerns about using FAIR Act inventory data to identify inherently governmental and core-commercial activities:

- *Improper classification of activities in the FAIR Act inventory.* Disagreements between the Forest Service and USDA regarding activity designations raise questions about appropriate classification. Furthermore, agency employees may be tempted to classify an activity as either inherently governmental or core-commercial to exempt it from competition.
- *Mandatory use of OMB function codes.* The Forest Service is required by OMB to use OMB-assigned function codes for the FAIR Act inventory. Although the guidance allows the Forest Service flexibility in defining the codes, the officials told us that some of the codes were too broad to be of any use or did not capture the actual work activities Forest Service employees carried out.[20]

Because Forest Service officials did not find the agency's FAIR Act inventory data useful during the A-76 precompetition planning stage, those officials responsible for the three completed competitions—IT infrastructure, road maintenance, and fleet maintenance—said that they developed their own methodologies to classify inherently governmental and core-commercial activities. For each competition, agency officials collected additional work activity information from field offices and performed additional analysis beyond that conducted for the FAIR Act inventory. In doing so, officials said, they received little guidance on how to supplement the FAIR Act inventory data or how to identify inherently governmental and core-commercial activities during the precompetition stage of the completed competitions.

Despite the lack of guidance, Forest Service officials involved in the three completed competitions said that they succeeded in identifying and exempting from competition inherently governmental and core-commercial activities by relying on intuitive knowledge and outside consultants. Officials were confident of their success because of their expertise in the activity being competed; the clearly commercial nature of the activities; and the small size of the competitions, especially the fleet maintenance and the road maintenance competitions. However, the three completed competitions may not be representative of future competitions.

Since these competitions, USDA and the Forest Service have taken steps to better define inherently governmental and core-commercial activities for the purposes of completing the FAIR Act inventory.[21] However, while we believe that the FAIR Act inventory data could be a useful tool for planning competitive sourcing activities at an agencywide level—for example, they can form the basis of the Green Plan—we believe that additional guidance will be needed on how to classify work activities during the precompetition planning stage of an A-76 competition to ensure that key work activities are excluded from the specific competition being planned. Without such guidance, the Forest Service is at risk of subjecting inherently governmental and core-commercial activities to A-76 competitions. This is particularly true as the Forest Service continues to implement its Green Plan, which could potentially subject up to two thirds of its FTEs to A-76 competitions.

The Forest Service Lacks a Strategy to Assess the Cumulative Effect That A-76 Competitions Could Have on Its Ability to Respond to Wildland Fires and Other Emergencies

For the three competitions the Forest Service completed—IT infrastructure, road maintenance, and fleet maintenance—officials responsible for planning the competitions told us they likely had a negligible effect on the Forest Service militia's ability to fight fires and respond to emergencies for the following reasons:

- The three competitions affected a relatively small number of Forest Service employees—1,323—compared with the approximately 80,000 federal employees who are ICS-certified.
- Contract provisions required the winning organization to provide emergency incident support for activities within the scope of the contract. For example, the MEO that won the road maintenance competition was obligated to provide road maintenance, if needed, to support the response to a wildfire incident.
- The largest of the three competitions—the IT infrastructure, which affected 1,200 FTEs—was won by the MEO. Because the MEO is still a unit of the Forest Service and staffed by Forest Service employees, the agency was able to direct the MEO to allow ICS-certified employees to volunteer to fight wildfires and respond to other emergencies by performing non-IT-related duties.

While the Forest Service has thus far minimized the impact of A-76 competitions on the availability of ICS-certified personnel to fight wildfires and respond to other emergencies, the following other factors may affect the availability of ICS-certified personnel in the future:

- The Forest Service cannot realistically expect a private sector firm to provide emergency services unrelated to the activity being competed. For example, the Forest Service could not hold a competition for fleet maintenance and expect firms that specialize in fleet maintenance to provide unrelated services at the scene of the fire, such as providing food and supplies.
- Whether an MEO or a private sector firm wins a competition, the availability of ICS-certified personnel could decline. As with any reorganization, competitive sourcing may cause some personnel to leave the Forest Service. These employees could retire or be hired by other federal agencies that participate in the NWCG, and thus they could continue to fight fires and respond to other emergencies.[22] Other employees, however, may no longer be available if their new employment situation does not allow them to take extended leaves of absence to fight fires and respond to other emergencies.

In its fiscal years 2006 and 2007 appropriations for the Forest Service, Congress required the agency, in carrying out any competitive sourcing competition involving Forest Service employees, to take into account the potential effect that contracting with a private sector organization would have on the agency's ability to fight and manage wildfires.[23] For the only A-76 competition started since the law was passed—the communications competition—

the 130 employees potentially affected by the competition were asked to report the amount of time they spent responding to emergencies during the previous year. However, the Forest Service did not collect information on what specific duties those employees performed during the emergency response, nor the ICS-qualifications they hold.

Without this information, the Forest Service cannot assess the full impact of this competition on its emergency response capability.

While it is important to know the impact of individual competitions, it is even more important to know the cumulative impact of multiple competitions. However, the Forest Service does not have a strategy to assess the cumulative impact that future competitions could have on its firefighting capability. The absence of such a strategy could prove significant if the Forest Service implements its plan to consider over 24,000 FTEs—or nearly two thirds of its workforce—for A-76 competition.

THE FOREST SERVICE DOES NOT HAVE SUFFICIENT COMPETITIVE SOURCING COST DATA TO DEMONSTRATE COMPLIANCE WITH ITS STATUTORY SPENDING LIMITATIONS OR TO ACCURATELY REPORT SAVINGS TO CONGRESS

The Forest Service does not know how much it spent on competitive sourcing activities and, therefore, cannot be assured that it stayed within the spending limitations or that it accurately reported savings to Congress. For fiscal years 2004 through 2006, we found that the Forest Service (1) narrowly interpreted the spending limitations to exclude certain costs and (2) lacked sufficiently complete and reliable cost data to demonstrate its compliance with the appropriations acts' spending limitations on its competitive sourcing activities. Furthermore, Congress may not have an accurate measure of the savings from the Forest Service's A-76 competitions because the agency (1) does not have complete and reliable cost data and (2) did not include all costs associated with its competitive sourcing program.

The Forest Service's Interpretation of Costs That Are Subject to Statutory Spending Limitations Was Too Narrow

For fiscal years 2004 through 2006, the Forest Service did not attempt to collect cost data on all competitive sourcing activities because it believed that some costs associated with these activities were not subject to the spending limitations of $5 million, $2 million, and $3 million, respectively, as established in its appropriations acts. Specifically, the Forest Service reasoned that it did not have to collect all the cost data because it interpreted the spending limitations as being intended to restrict the number of A-76 competitions it conducted. It, therefore, asserted that the costs associated with the competition stage were subject to the spending limitations, while costs associated with the FAIR Act inventory; precompetition planning; and postcompetition accountability activities should not be included.

However, we found that the Forest Service's interpretation of which competitive sourcing activities are subject to the spending limitations was too narrow. Specifically, we concluded that, with only a limited exception, the spending limitations apply to all costs attributable to the Forest Service's competitive sourcing program, including feasibility studies and other precompetition planning activities, the competition itself, postcompetition accountability activities, and the CSPO's costs to manage the program. Only the costs incurred to comply with the FAIR Act, such as those to develop the inventories of activities, are exempt from the limitations because the Forest Service is statutorily required to perform FAIR Act-related activities even if it makes no effort to conduct competitive sourcing.[24] (See app. II for further discussion on our legal interpretation.)

In April 2007, we sought the opinion of the USDA's General Counsel on whether certain Forest Service competitive sourcing activities are subject to the annual statutory spending limitations. Our interpretation of the costs that were subject to the spending limitations for fiscal years 2004 through 2006 is consistent with the interpretation that USDA's General Counsel provided to us. The USDA's General Counsel further stated that it believed that the Forest Service may not have complied with the spending limitations in fiscal years 2004 through 2006.

The Forest Service Did Not Ensure That Cost Data Used to Comply with Statutory Spending Limitations Are Reliable

Even when it used its own interpretation of costs subject to the spending limitations, the Forest Service still did not know whether it complied with the limitations because it did not have a cost accounting system sufficient to track costs related to competitive sourcing. First, the Forest Service failed to establish tracking codes, known as job codes, in its financial management system to enable it to distinguish cost data on the activities that it believed were subject to the spending limitations from other cost data.[25] Forest Service officials could not explain why the agency had not established these job codes.

Second, the Forest Service lacked guidance and management oversight to ensure that employees were accurately and consistently using the job codes that were established for competitive sourcing activities. In particular, Forest Service officials could not provide us with any guidance that employees could use to determine when to charge time to these job codes and when to charge time to codes associated with their regular duties. Officials acknowledged that without this guidance, employees probably continued to charge time spent on competitive sourcing activities to their regular job codes. For example:

- In fiscal year 2004, Forest Service employees charged only 0.07 FTEs to the job code established to track costs with the IT infrastructure competition, even though competitive sourcing activities for the competition took place throughout the entire fiscal year.
- In fiscal years 2005 through 2006, Forest Service employees charged only 0.22 FTEs to the job code established to track costs associated with the communications competition, even though competitive sourcing activities for the competition began in fiscal year 2005 and were ongoing at the end of fiscal year 2006.

In consultation with Forest Service officials, we agreed that it was not feasible to reconstruct cost data for competitive sourcing activities between fiscal years 2004 and 2006 to determine if the Forest Service exceeded the appropriations acts' spending limitations.

Forest Service officials told us it would require a significant amount of time and resources to query employees on their past work activities associated with competitive sourcing. Furthermore, it is unlikely that employees could reliably report the time they spent on competitive sourcing activities that took place months and years ago. Finally, officials told us that many employees involved with the competitions have since left the agency.

Recognizing these shortcomings, the Forest Service has made some efforts to improve its policies and guidance on how to establish job codes and how employees are to use them to track competitive sourcing costs. In fiscal year 2007, the Forest Service issued general policy on when to charge time to competitive sourcing job codes. Among other things, it issued a directive specifying that the cost of performing some precompetition planning activities be charged to competitive sourcing job codes. However, the Forest Service has yet to provide details on how it intends to implement this policy, and thus we were not able to evaluate it.

The Forest Service Could Not Substantiate Savings Reported to Congress

The Consolidated Appropriations Act, 2004, establishes a governmentwide requirement for each executive agency to report to Congress on the actual savings derived from the implementation of competitions for the prior fiscal year.[26] In addition, OMB provides guidance on preparing the report, including how to calculate savings.[27] According to the guidance, savings from completed competitions is defined as the difference between the cost to the federal government of performing the activity prior to the competition—the baseline cost—and the cost to the government of performing the activity or paying for it after the winner of the competition has begun performing the activity. This is the postcompetition cost. Figure 1 shows how savings from competitions are calculated.

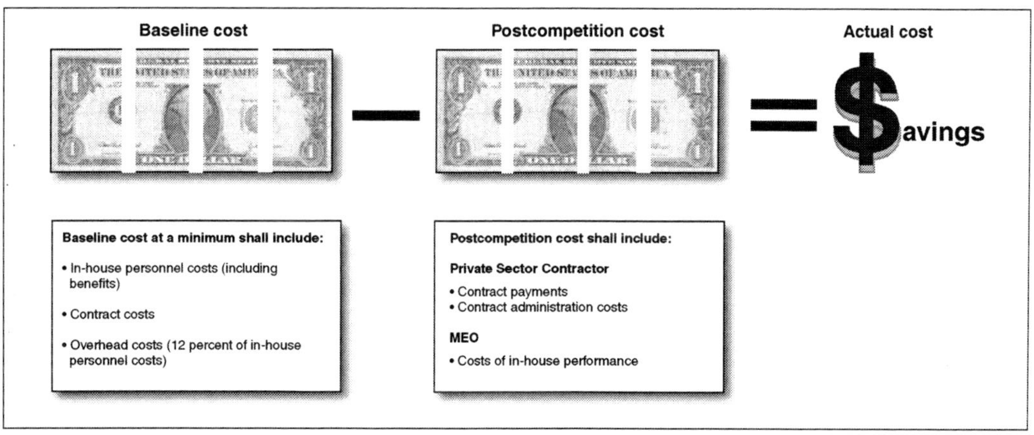

Source: GAO representation of OMB guidance.

Figure 1. OMB Calculation for Determining Savings from Competitions, Fiscal Years 2004-2006.

While OMB guidance specifies the postcompetition cost to the government if a private sector contractor wins the competition, it does not provide specific direction on how to calculate the postcompetition cost to the government if the MEO wins the competition. Instead, the guidance suggests determining the cost of in-house performance using a methodology similar to that used to calculate the baseline costs.

In its reports to Congress on the actual savings derived from the implementation of competitions, USDA reported that the Forest Service saved over $38 million between fiscal years 2004 through 2006 as a result of the three completed competitions—IT infrastructure (approximately $35.2 million), fleet maintenance (approximately $716,000), and road maintenance (approximately $2.2 million).

For these three competitions, Forest Service officials at the agency's headquarters and in the regions could not provide us with the information necessary to fully substantiate the savings reported to Congress. Headquarters officials directed us to the regions responsible for the competitions, where officials were able to provide some cost data, but overall could not provide us with the information necessary to substantiate the savings reported to Congress. GAO's document entitled

The Standards for Internal Control in the Federal Government requires that all transactions and other significant events, such as the Forest Service competitive sourcing costs and savings analyses, be clearly documented, and the documentation should be readily available for examination.[28] Specifically, we found the following:

- *IT infrastructure.* Officials could not tell us the methodology they used to determine postcompetition costs and did not have available the data they used to calculate savings. They reconstructed the personnel and overhead costs for operating the MEO, but could not reconstruct the cost of outside contractors that the MEO employed because these costs were not discernable from other cost data that the Forest Service collected. Ultimately, these officials could only speculate about what the contract costs might have been and how savings were calculated.
- *Fleet maintenance.* As with the IT infrastructure competition, officials could not tell us the methodology they used to determine postcompetition costs and did not have available all of the data they used to calculate savings. Although officials provided us with the bulk of the postcompetition cost data—the payments made to Serco, the private sector organization that won the competition—they could not provide us with all of the costs.
- *Road maintenance.* Unlike the other two competitions, officials responsible for this competition described the methodology. However, they were unable to provide us with baseline cost data for fiscal year 2004, and, as a result, we could not verify the reported savings for that year.

In April 2007, OMB issued guidance to help agencies substantiate the savings they have achieved through A-76 competitions. The guidance describes agency responsibilities related to tracking and reviewing cost data to ensure savings are being realized. It also requires all agencies to develop plans to independently validate a sampling of competitions to confirm projected savings. Among other things, the guidance requires agencies to assess the completeness and accuracy of cost data. Forest Service officials told us that the agency is working to implement OMB's guidance. USDA told us that it is directing the Forest Service to validate the savings from the IT infrastructure competition by the fourth quarter of fiscal year 2008.

The Forest Service Did Not Include All Costs Associated with Its Competitive Sourcing Program When Calculating Savings Reported to Congress

While the Forest Service could not substantiate the savings it reported to Congress using OMB's guidance, the guidance itself allows agencies to exclude some costs associated with A-76 competitions, which, if excluded, may not provide Congress with an accurate measure of the savings produced by the competitions. Although Forest Service officials could not tell us all of the costs that were included or excluded in their savings estimates, they stated with confidence that some costs were excluded. Specifically, they said that transition costs associated with transferring responsibilities to the winning organization were not included in the savings calculations for the three competitions we reviewed. Transition costs include costs associated with decreasing the size of the workforce through buyouts and retirements, and costs to transfer employees who are being retained to other locations. These are not necessarily one-time costs because an agency is required to complete a follow-on competition for an activity, generally after about 3 to 5 years. In the three competitions we reviewed, we found the following transition costs were excluded:

- *IT infrastructure.* Forest Service officials told us that they excluded approximately $40 million in transition costs from the savings calculations that were reported to Congress.[29] These costs exceeded by about $5 million the $35.2 million that the Forest Service reported to have saved during fiscal years 2005 and 2006 as a result of the competition.
- *Fleet maintenance.* Forest Service officials told us they excluded about $670,000 in transition costs from the savings calculations that were reported to Congress, nearly equaling the approximately $716,000 that the Forest Service reported to have saved since fiscal year 2005 as a result of the competition.
- *Road maintenance.* Forest Service officials told us they excluded about $320,000 in transition costs from the savings calculations that were reported to Congress. These costs are approximately 15 percent of the total $2.2 million in reported savings since fiscal year 2004.

In addition to transition costs, there are also precompetition planning costs (including, as of May 2004, feasibility study costs). Like transition costs, OMB guidance does not direct that precompetition planning costs be included in the savings calculations. A Forest Service official told us that precompetition planning costs were excluded from the savings calculations for the three completed competitions. Forest Service officials could not provide us with estimates of these costs because they were not tracked.

OMB guidance also does not direct agencies to include in their savings calculation other potential costs associated with the termination of a contract. To illustrate, 14 months after the private sector contractor, Serco, began performing fleet maintenance activities, the Forest Service terminated the contract. Under the terms of the contract, the Forest Service must negotiate with Serco on any costs associated with the termination itself. Forest Service officials stated that in March 2007, Serco proposed a settlement amount that it contends would reimburse it for costs such as those associated with vacating sites as well as administrative costs and attorneys' fees. Serco also contends that the Forest Service owes it additional compensation unrelated to the contract termination. Serco's claim stems from a disagreement with the Forest Service over the terms of the contract. Finally, Forest Service officials also told us that the agency has incurred additional costs—exceeding its costs for fleet maintenance before the Serco contract—because maintenance work is now being performed by retail venders. Officials told us that the Forest Service is precluded from returning the work to the agency for in-house performance.[30] Because these issues are still pending, we did not include the dollar amounts of Serco's claims or the additional costs incurred by the Forest Service, and the Forest Service officials told us that they did not want to comment further on this issue.

CONCLUSIONS

The government's goal is to obtain high-quality services at a reasonable cost, regardless of whether these services are performed by the public or private sector. Competition between the public and private sectors is an important tool in reaching this goal. The 2001 *President's Management Agenda* has reemphasized the importance of using this tool as a means to deliver the best value to the American taxpayer.

In keeping with the *President's Management Agenda*, the Forest Service established a competitive sourcing program and began holding A-76 competitions. The agency now has plans to consider competing up to two thirds of its workforce against the private sector. This is a massive undertaking whose long-term success will depend on a realistic strategic plan, clear guidance to identify the key work activities that should be excluded from competition, and a strategy to assess the cumulative effect that outsourcing a large number of federal jobs could have on its firefighting capability. Unfortunately, the Forest Service has none of these in place. Because the Forest Service has just begun to implement its competitive sourcing program—having competed less than 5 percent of its workforce against the private sector—these problems have not yet had a significant impact on the Forest Service's competitive sourcing program. However, as the Forest Service implements its Green Plan and greatly expands the scope of its competitive sourcing program, the problems we have identified could, in the long term, severely impact its ability to implement this program effectively—jeopardizing not just the overall success of the program, but the nation's ability to fight fires and respond to other emergencies.

More immediate is our concern that the Forest Service did not collect complete and reliable cost data related to its competitive sourcing program during fiscal years 2004 through 2006. As a result, the agency did not know how much it had spent on competitive sourcing activities and, consequently, whether it had complied with statutory spending limitations. We are also concerned about the usefulness of the cost savings that the Forest Service reported to Congress. The Forest Service could not provide us with sufficient data to verify the accuracy of the reported savings and excluded substantial costs from the savings calculations. Although the agency followed OMB guidance in calculating savings, we believe the guidance provides the Forest Service with the latitude to include the other costs we identified—some of which were substantial. Including these costs would have provided Congress with a more realistic picture of the extent to which the Forest Service's competitive sourcing program is saving the American taxpayers' money.

APPENDIX I: OBJECTIVES, SCOPE, AND METHODOLOGY

This report discusses the extent to which the U.S. Department of Agriculture's (USDA) Forest Service has (1) plans and guidance to help implement its competitive sourcing program effectively and (2) sufficient cost data to ensure that it complied with its competitive sourcing statutory spending limitations and accurately reported its competitive sourcing savings to Congress for fiscal years 2004 through 2006.

To address these two objectives, we focused on Forest Service competitive sourcing activities between fiscal years 2004 and 2006. For the first objective, we compared the Forest Service's Green Plan, including draft versions of the plan, with previous work that GAO has performed that identified the elements of effective strategic plans. We also reviewed the guidance that the Office of Management and Budget (OMB) and USDA provided to the Forest Service on how to construct its Green Plan. Furthermore, we examined how the Forest Service scoped the three competitions that it completed during this time frame. Specifically, we examined how the Forest Service ensured that it did not include inherently governmental and core-commercial activities in the competitions, and how it assessed the impact that competitions could have on the nation's ability to fight wildland fires and respond to other emergencies. Finally, we examined the guidance currently available to Forest Service employees responsible for scoping competitive sourcing competitions and interviewed Forest Service officials familiar with the agency's competitive sourcing activities.

To respond to the second objective, we examined the appropriations acts for fiscal years 2004 through 2006 to determine which costs incurred by the Forest Service during those fiscal years were subject to the spending limitations. Appendix II explains this analysis in greater detail. We then asked the Forest Service to provide us with cost data for activities associated with its competitive sourcing program, including cost data we had determined were subject to the spending limitations. We assessed the reliability of these data and found them unreliable as a measure of the costs and savings associated with the program. Specifically, our review of cost data from the USDA's Federal Financial Information System (FFIS) and subsequent discussions with Forest Service officials, including the Chief Financial Officer, confirmed that the Forest Service had not issued formal guidance on which costs where to be charged to competitive sourcing job codes in FFIS, and that there was a lack of institutionalized policies and procedures to guide employees in tracking the time and costs associated with competitive sourcing activities. As a result, Forest Service officials, including the Chief Financial Officer, agreed that it was likely that employees used job codes inconsistently and that the accounting of competitive sourcing expenditures between fiscal years 2004 and 2006 were unreliable and inaccurate. We also asked the Forest Service for data to substantiate the annual savings resulting from competitive sourcing competitions it reported to Congress in fiscal years 2004 through 2006. For the three competitions completed during this time frame, Forest Service officials either could not provide us with the methodology used to calculate the savings, or could not provide us with all of the data used to calculate the savings. As a result, we were unable to replicate the Forest Service's savings calculations, and thus we determined the reported numbers were unreliable because data reporting strategies were absent or inconsistent, there were no internal processes for verifying the accuracy of the data, and there was no documentation with which we could verify the accuracy of the data. Finally, in the course of our work, we identified categories of costs that were not included in the Forest Service's savings calculations. We include examples of these costs in this report to show their magnitude relative to the savings reported to Congress; however, we did not independently verify the accuracy of these costs. We conducted our work between October 2006 and January 2008 in accordance with generally accepted government auditing standards.

APPENDIX II: LIMITATIONS ON USE OF APPROPRIATIONS FOR "COMPETITIVE SOURCING STUDIES AND RELATED ACTIVITIES"

The Forest Service relies on both federal employees and private contractors to perform commercial activities—activities performed by the government that are also regularly performed in the commercial market place—such as data entry, photocopying, and other administrative support services. To determine whether to convert the performance of such activities from federal employees to private contractors, or *vice versa*, the Forest Service follows the competitive process set out in Office of Management and Budget (OMB) Circular No. A-76, *Performance of Commercial Activities* (May 29, 2003), as revised, which establishes executive branch policy for the competition of commercial activities.[1] The Forest Service is also subject to the requirements of the Federal Activities Inventory Reform Act of 1998 (FAIR Act),[2] which mandates agencies to develop inventories of their activities, and the President's Management Agenda (PMA), launched in 2001, which includes a government-wide "competitive sourcing" initiative to increase and to make more efficient and effective agencies' use of the circular's competitive process.[3] To implement these requirements, the Forest Service established a competitive sourcing program that is administered by the Competitive Sourcing Program Office.[4]

A provision in each of the annual appropriations acts funding the Forest Service for fiscal years 2004 through 2006 limited the amount of appropriated funds available to the Forest Service for "competitive sourcing studies and related activities" to between $2 million and $5 million (spending limitations). Five members of the U. S. Senate requested that we determine which costs incurred by the Forest Service in fiscal years 2004 through 2006 were subject to the spending limitations. In our view, these spending limitations apply to all costs under the Forest Service's competitive sourcing program, except for costs incurred to carry out the FAIR Act.

Background

OMB Circular No. A-76 includes, in appendix B, detailed guidance on the competitive process a federal agency should follow to determine whether an activity should be performed by a public or private source.[5] The process is divided into three stages: (1) a precompetition preliminary planning stage, consisting, at minimum, of nine specified steps; (2) a competition stage that starts with a formal public announcement of a standard or streamlined competition[6] between private and public sources and ends with announcing a performance decision; and (3) a postcompetition accountability stage, which mandates such actions as transferring performance of the activities to the winning entity, maintaining a database to track the execution of competitions, monitoring performance, and submitting reports to OMB.

To avoid the mistakes it made during earlier competitions, the Forest Service has supplemented this process since 2004 by first conducting studies to determine the feasibility of carrying out this competitive process (feasibility studies) for each activity.[7] In 1998, Congress enacted the FAIR Act, which requires federal agencies to submit to OMB their annual inventories of commercial activities and, after a period of review and consultation with OMB, transmit the final inventories to Congress and make them available to the public. Appendix A to OMB Circular No. A-76 (*Inventory Process*) implements the FAIR Act.

In August 2001, the Administration launched the PMA as its reform agenda for improving management and performance in the federal government. The PMA includes a competitive sourcing initiative to simplify and improve the process for choosing private or public sources. To implement the PMA competitive sourcing initiative, the Forest Service stated in its fiscal year 2004 budget justification that it planned to conduct competitions for activities associated with a total of 11,000 full-time employees or their equivalent (FTEs), and that competitive sourcing costs were expected to rise.[8] Concerned about how the Forest Service had implemented the PMA competitive sourcing initiative,[9] the House and Senate appropriations committees included language in the appropriations bills for fiscal year 2004 that led to the enactment of the following spending limitation: "Of the funds appropriated by this Act, not more than $5,000,000 may be used in fiscal year 2004 for competitive sourcing studies and related activities by the Forest Service."[10] The Forest Service's appropriations acts for fiscal years 2005 and 2006 included spending limitations with identical language, except that the maximum amount available for this purpose was $2 million and $3 million, respectively.[11] At issue here is determining the range of Forest Service activities that are covered by the appropriations acts' spending limitations on "competitive sourcing studies and related activities."

Discussion

Construing the meaning of a statutory provision starts with the statutory language.[12] The appropriations acts for fiscal years 2004 through 2006 define a "competitive sourcing study" for purposes of the spending limitations as

> "a study on subjecting work performed by Federal Government employees or private contractors to public-private competition or on converting the Federal Government employees or the work performed by such employees to private contractor performance under the Office of Management and Budget Circular A–76 or any other administrative regulation, directive, or policy."[13]

Unfortunately, by defining a "competitive sourcing study" as a "study," albeit in the context of private versus public source performance,[14] the appropriations acts do not clearly indicate the breadth of Forest Service activities that fall within its meaning.[15] At the very minimum, however, the phrase includes the competition stage (the second stage) of the OMB Circular No. A-76 process because it represents the formal structured process of soliciting and evaluating proposals from public and private sources. The Forest Service takes the position that the annual spending limitations apply only to costs attributable to the competition stage.[16] We disagree, as does the U. S. Department of Agriculture's General Counsel, who provides legal assistance to the Forest Service.[17] Even if we were to read the word "study" as narrowly as the Forest Service apparently does, the additional words "and related activities" in this context make clear that the statutory language includes more than simply the competitions.

The statutes do not define "related activities" or set out specific criteria for identifying them. While a broad range of activities are "related," in the sense of having some logical connection to competitive sourcing studies, we believe Congress intended to focus on those activities that the Forest Service performs as part of its implementation of the President's competitive sourcing initiative.[18] In other words, "related activities" are those activities the Forest Service performs if it is considering whether to conduct a public-private competition. Under the Forest Service's competitive sourcing program, no competition is conducted until the completion of a feasibility study, the objective of which is to evaluate whether an activity should be subject to the A-76 competitive sourcing process. Likewise, no competition is conducted under OMB Circular No. A-76 until the completion of the precompetition stage (the first stage), which also includes tasks such as conducting market research and developing an acquisition plan. Feasibility studies and precompetition planning activities are thus prerequisites to a competition and, therefore, in our view, "related activities." Similarly, the postcompetition stage is a required consequence of conducting a competition. The Forest Service's Competitive Sourcing Program Office supports and manages these and other activities in the Forest Service's competitive sourcing program. Because these activities further the Forest Service's objective of conducting competitions under the competitive sourcing initiative, we view them as "related activities" subject to the spending limitation.

While it can be argued that the words "and related activities" include activities the Forest Service performs to carry out the FAIR Act, such as preparing inventories and reporting to Congress, we think the better view is that these costs are not encompassed in this limitation. Congress by law expressly requires these activities, even absent any other competitive sourcing program activities. Moreover, complying with the FAIR Act does not further the Forest Service's progress on the PMA competitive sourcing initiative, and it was the Forest Service's cost to implement this initiative to which the congressional committees were reacting.[19] Our interpretation is consistent with the interpretation provided to us by the USDA Office of General Counsel.[20]

Conclusion

Accordingly, we conclude that the spending limitations on "competitive sourcing studies and related activities" included in the appropriations acts funding the Forest Service for fiscal years 2004 through 2006 apply to all costs attributable to the Forest Service's competitive sourcing program, including the following:

- conducting feasibility studies to evaluate which activities should be subject to the competitive sourcing process set out in OMB Circular No. A-76;
- engaging in the competitive sourcing process set out in OMB Circular No. A-76, including activities related to precompetition planning, competitions between sources, and postcompetition accountability; and
- managing the agency's competitive sourcing program.

The appropriations acts spending limitations, however, do not apply to costs incurred for complying with the FAIR Act, which the Forest Service must do even if it applies no efforts to competitive sourcing.

REFERENCES

[1] The Executive Office of the President and the Office of Management and Budget, *The President's Management Agenda* (2001).

[2] OMB Circular No. A-76, *Performance of Commercial Activities* (May 29, 2003), and Pub. L. No. 105-270, 112 Stat. 2382 (Oct. 19, 1998).

[3] Pub. L. No. 108-199, Div. F, Title VI, § 647(b), 118 Stat. 3, 361 (Jan. 23, 2004).

[4] OMB memoranda, M-07-01 (Oct. 5, 2006); M-06-01 (Oct. 7, 2005); and M-05-01 (Oct. 15, 2004).

[5] Department of the Interior and Related Agencies Appropriations Act, 2004, Pub. L. No. 108-108, § 340(d)(3), 117 Stat. 1241, 1316 (Nov. 10, 2003); Department of the Interior and Related Agencies Appropriations Act, 2005, Pub. L. No. 108-447, div. E, § 332(a)(3), 118 Stat. 2809, 3100 (Dec. 8, 2004); Department of the Interior, Environment, and Related Agencies Appropriations Act, 2006, Pub. L. No. 109-54, § 422(a)(2), 119 Stat. 499, 554 (Aug. 2, 2005); and Revised Continuing Appropriations Resolution, 2007, Pub. L. No. 110-5, § 104, 121 Stat. 8, 9 (Feb. 15, 2007) (operates to retain the spending limitation imposed on the Forest Service in the fiscal year 2006 appropriations act).

[6] USDA, Forest Service, Report to Surveys and Investigations Staff, Committee on Appropriations, U.S. House of Representatives, *Implementation of the Competitive Sourcing Initiative at the U.S. Forest Service* (March 2004).

[7] The Geospatial Service and Technology Center is a Forest Service facility primarily located in Salt Lake City, Utah, that provides geographic data and mapping services to all departments within the Forest Service.

[8] There was a fourth competition completed during this time frame that considered 8 FTEs. We excluded this competition from our analysis because it was conducted using the streamlined competition process, which is faster and less complex than the standard competition process that was used for the other three competitions.

[9] The NWCG draws on employees from the Forest Service and from the Department of the Interior's Bureau of Land Management, Bureau of Indian Affairs, Fish and Wildlife Service, and National Park Service.

[10] OMB Circular No. A-76, *Performance of Commercial Activities* (May 29, 2003).

[11] See *The President's Management Agenda* (2001). The five initiatives are strategic management of human capital, improved financial performance, expanded electronic government, budget and performance integration, and competitive sourcing.

[12] The FAIR Act, Pub. L. No. 105-270, § 5, 112 Stat. 2382, 2384 (Oct. 19, 1998).

[13] OMB guidance for use of commercial exemption justifications states the following: "As a general matter, a function should be considered core to an agency's operation only if—and only to the extent that—loss of in-house performance of the function would result in substantial risk to the agency's ability to accomplish its unique mission." See OMB, *2005 Inventories of Commercial and Inherently Governmental Activities,* M-05-12 (May 23, 2005).

[14] Besides core-commercial, there are four other subcategories of commercial activities that are exempt from competition. Commercial activities are exempt if they are (1) subjects of an in-progress A-76 competition, (2) performed by government personnel as the result of an A-76 competition within the past 5 years, (3) involved in agency restructuring, or (4) performed by government personnel due to a statutory prohibition against private sector performance.

[15] The name Green Plan is derived from the July 2003 OMB standards for evaluating federal agencies' progress in implementing PMA initiatives. These standards describe the degree of each agency's progress as red, yellow, or green, where green is considered fully successful. One OMB standard requires federal agencies to develop a competition plan as a requirement to earn green status for the competitive sourcing initiative.

[16] See footnote 5.

[17] Commercial B activities are commercial activities suitable for competition—they exclude inherently governmental and core-commercial activities.

[18] GAO, *Congressional Review of Agency Strategic Plans*, GAO/GGD-10.1.16 (Washington, D.C.: May 1997); *Transforming the Civil Service: Building the Workforce Of the Future, Results Of a GAO-Sponsored Symposium*, GAO/GGD-96-35 (Washington, D.C.: Dec. 20, 1995); *Managing for Results: Experiences Abroad Suggest Insights for Federal Management Reform*, GAO/GGD-95-120 (Washington, D.C.: May 2, 1995); *Managing for Results: State Experiences Provide Insights for Federal Management Reforms*,

[19] GAO/GGD-95-22 (Washington, D.C.: Dec. 21, 1994); *Government Reform: Goal-Setting and Performance*, GAO/AIMD/GGD-95-130R (Washington, D.C.: Mar. 27, 1995); and *Executive Guide: Improving Mission Performance Through Strategic Information Management and Technology*, GAO/AIMD-94-115 (Washington, D.C.: May 1994).

[20] GAO, *Competitive Sourcing: Greater Emphasis Needed on Increasing Efficiency and Improving Performance*, GAO-04-367 (Washington, D.C.: Feb. 27, 2004).

[21] As we have previously reported, the inapplicability of the OMB-assigned codes has been a problem for many other federal agencies. Our prior work on this issue includes GAO-04-367 and *Competitive Contracting: The Understandability of FAIR Act Inventories Was Limited*, GAO/GGD-00-68 (Washington, D.C.: Apr. 14, 2000).

[22] See, for example, U.S. Department of Agriculture, Office of the Chief Financial Officer, Office of Competitive Sourcing, *FY07 Guidebook for Inventories of Commercial Activities and Inherently Governmental Activities* (March 2007).

[23] Retired employees who are ICS-certified are eligible to be hired on an hourly basis to fight fires and respond to other emergencies.

[24] Pub. L. No. 109-54 § 422(e), 119 Stat. 499, 554-55 (Aug. 2, 2005); and Pub. L. No. 110-5, § 104, 121 Stat. 8, 9 (Feb. 15, 2007) (operates to retain this requirement in fiscal year 2007). Section 422(e) states the following: "determine whether any of the employees concerned are also qualified to participate in wildland fire management activities; and take into consideration the effect that contracting with a private sector source would have on the ability of the Forest Service to effectively and efficiently fight and manage wildfires."

[25] For fiscal year 2008, the Forest Service is prohibited from using any fiscal year 2008 funds for "competitive sourcing studies and related activities involving Forest Service

personnel." Consolidated Appropriations Act, 2008, Div. F, § 415(a), Pub. L. No. 110-161 (Dec. 26, 2007).

[26] The Forest Service uses job codes to track its competitive sourcing program costs, which are established in its Foundation Financial Information System. The job codes are used to track costs related to things such as salary, travel, consultancy services, or supplies.

[27] Specifically, the Consolidated Appropriations Act, 2004, requires agencies to report on "actual savings, or a quantifiable description of improvements in service or performance, derived from the implementation of competitions" for the previous fiscal year. Pub. L. No. 108-199, Div. F, Title VI, § 647(b)(7), 118 Stat. 3, 361 (Jan. 23, 2004).

[28] OMB provides this guidance annually. The guidance did not change substantially for fiscal years 2004 through 2006. See footnote 4.

[29] GAO, *The Standards for Internal Control in the Federal Government*, GAO/AIMD-00-21.3.1 (Washington, D.C.: November 1999).

[30] The Forest Service reported that the savings accrued over a 20-month period between February 2005 and September 2006.

[31] According to the circular, work performed by a commercial source cannot automatically revert to government (in-house) performance. Specifically, before the government may perform the work currently being performed by the private sector, an A-76 competition must be used to determine whether the government should perform the activity.

Appendix II

[1] OMB Circular No. A-76 may be accessed at www.whitehouse.gov/omb/circulars/index.html (last visited Oct. 22, 2007).

[2] Pub. L. No. 105-270, 112 Stat. 2382 (Oct. 19, 1998).

[3] Executive Office of the President and OMB, *The President's Management Agenda* (2001), *available at* www.whitehouse.gov/results/agenda/index.html (last visited Sept. 6, 2007).

[4] USDA-Forest Service, Report to Surveys and Investigations Staff, Committee on Appropriations, U.S. House of Representatives, *Implementation of the Competitive Sourcing Initiative at the U.S. Forest Service,* at i, 3, 4, 9 (Mar. 2004).

[5] The circular was first issued in 1966 and last revised in 2003. *See* OMB Cir. No. A-76, ¶ 4 (June 14, 1999), ¶ 2 (May 29, 2003).

[6] When 65 or fewer full-time employees or its equivalent (FTEs) are required to perform the commercial activity being evaluated, OMB Circular No. A-76 permits a federal agency to perform a streamlined competition, which is faster and less costly to conduct than a standard (*i.e.,* regular) competition because of the abbreviated and less complex procedures required. Under older versions of the circular, the "competition" held under stage two was called a "cost comparison." *See* OMB Cir. No. A-76, ¶ 7; app. D at D-3 (definition for "competition").

[7] USDA OCFO, *Guidance for Determining the Feasibility of Conducting Competitive Sourcing Competitions,* Bulletin 2004-001 (May 11, 2004). The bulletin lists 13 categories of information that are needed to make this determination, such as Business Needs Assessment, Market Research, Cost/Benefit Analysis, Acquisition Strategy, and Requirements.

[8] USDA Forest Service–FY 2004 Budget Justification, at 1–6 (Feb. 3, 2003), *available at* www.fs.fed.us/aboutus/budget/ (last visited Oct. 22, 2007).

[9] H.R. Rep. No. 108-195, at 107 (2003); S. Rep. No. 108-89, at 8 (2003).

[10] Department of the Interior and Related Agencies Appropriations Act, 2004, Pub. L. No. 108-108, § 340(d)(3), 117 Stat. 1241, 1316 (Nov. 10, 2003).

[11] Department of the Interior and Related Agencies Appropriations Act, 2005, Pub. L. No. 108-447, div. E, § 332(a)(3), 118 Stat. 2809, 3100 (Dec. 8, 2004); Department of the Interior, Environment, and Related Agencies Appropriations Act, 2006, Pub. L. No. 109-54, § 422(a)(2), 119 Stat. 499, 554 (Aug. 2, 2005).

[12] As the U.S. Supreme Court has stated: "There is, of course, no more persuasive evidence of the purpose of a statute than the words by which the legislature undertook to give expression to its wishes." *United States v. American Trucking Associations, Inc.*, 310 U.S. 534, 543 (1940); *see Mallard v. U.S. District Court*, 490 U.S. 296, 300–02 (1989); B-302973, Oct. 6, 2004; B-288173, June 13, 2002.

[13] Pub. L. No. 108-108, § 340(f); Pub. L. No. 108-447, § 332(b); Pub. L. No. 109-54, § 422(b).

[14] The term "competitive sourcing" originated with the Department of Defense in the late 1990s. GAO, *DOD Competitive Sourcing: Results of Recent Competitions,* GAO/NSIAD-99-44 (Washington, D.C.: Feb. 23, 1999), at 2. Its use in connection with the spending limitations, however, appears to have its origins in the 2001 PMA. Prior to 2001, the term "competitive sourcing" does not appear within the commercial activities executive guidance, such as the 1999 version of OMB Circular No. A-76, or statutes, such as the FAIR Act. However, the term appears in related executive guidance and statutes after the launch of the PMA, such as the 2003 version of OMB Circular No. A-76 (*e.g.,* "competitive sourcing quarterly reports" and "competitive sourcing officials") and the Consolidated Appropriations Act, 2004, Pub. L. No. 108-199, § 647(b), 118 Stat. 3, 361 (Jan. 23, 2004) (*e.g.,* "report on the competitive sourcing activities" of the agency).

[15] For example, in the context of OMB Circular No. A-76, a study could reasonably include or exclude the postcompetition stage activities, such as performance monitoring and reporting. In this regard, the circular appears to include them in its description of the competitive sourcing process. *See* OMB Cir. No. A-76, app. B, fig. B1, at B-4; fig. B2, at B-6. On the other hand, the spending limitations' statutory definition of a "competitive sourcing study" is similar to the definition of a "competition" under the current version of OMB Circular No. A-76, or a "cost comparison" under older versions. The current version of OMB Circular No. A-76 defines a "competition" as a "formal evaluation of sources to provide a commercial activity that uses pre-established rules (*e.g.,* the FAR, this circular) . . . includ[ing] streamlined and standard competitions performed in accordance with this circular." The supplement to the 1999 version of OMB Circular No. A-76 defined a "cost comparison" for the "activity under study" as "the process whereby the estimated cost of Government performance of a commercial activity is formally compared, in accordance with the principles and procedures of this Circular and Supplement, to the cost of performance by commercial or ISSA [interservice support agreement] sources."

[16] Memorandum from Chief Dale N. Bosworth, Forest Service, to Charles Christopherson, CFO, USDA, *Revised USDA Forest Service Competitive Sourcing "Green Plan": FY*

2005-2009, Dec. 23, 2005; Letter from L. Benjamin Young, Jr., Assistant General Counsel, USDA, to F. Abe Dymond, Assistant General Counsel, GAO, June 14, 2007 (Young Letter).

[17] Young Letter at 3. Because the Forest Service, a USDA agency, receives legal assistance from the USDA Office of General Counsel (OGC), we solicited the General Counsel's legal views.

[18] For example, in 2004 the House committee noted its concern about how the Forest Service had implemented the competitive sourcing "initiative" because it "has looked into this issue in detail and found a number of cases of mismanagement of this effort" and thus addressed "this issue" in bill language by "limiting the use of funds for competitive sourcing efforts." H.R. Rep. No. 108-542, at 111 (2004). The fiscal year 2004 conference committee report stated that the committee was concerned about the failure to budget adequately for the cost of implementing the competitive sourcing initiative and that "significant sums" were being expended on the competitive sourcing initiative "at the expense of critical, on-the-ground work." H.R. Conf. Rep. No. 108-330, at 85–86 (2003); *see also* H.R. Rep. No. 108-195, at 8–10.

[19] H.R. Rep. No. 108-195, at 8–10, 107. According to information the Forest Service provided to GAO, the cost of complying with the FAIR Act is negligible compared to the costs of conducting the other activities associated with the competitive sourcing program.

[20] According to the USDA OGC, the spending limitations do not apply to activities required by the FAIR Act but do apply to competitions, postcompetition accountability activities, and "all activities leading up to a determination of whether a competition of a commercial activity identified by a FAIR Act inventory should be conducted." Young Letter at 3.

Chapter 3

WILDLAND FIRE MANAGEMENT: FEDERAL AGENCIES LACK KEY LONG- AND SHORT-TERM MANAGEMENT STRATEGIES FOR USING PROGRAM FUNDS EFFECTIVELY *

WHY GAO DID THIS STUDY

The nation's wildland fire problems have worsened over the past decade. Recent years have seen dramatic increases in the number of acres burned and the dollars spent on preparing for and responding to wildland fires. As GAO has previously reported, a number of factors have contributed to worsening fire seasons and increased firefighting expenditures, including an accumulation of fuels due to past land management practices; drought and other stresses, in part related to climate change; and an increase in human development in or near wildlands.

Recent GAO reports have identified shortcomings in the approach to wildland fire management taken by the responsible federal agencies— the Department of Agriculture's Forest Service and four agencies within the Department of the Interior. GAO was asked to testify on agency efforts to (1) develop a cohesive strategy for preparing for and responding to wildland fire, (2) contain federal expenditures related to wildland fire, and (3) improve the processes used to allocate funds for reducing accumulated fuels and to select fuel reduction projects. GAO also is providing preliminary findings from its ongoing review of an interagency budget allocation and planning model known as fire program analysis (FPA). This testimony is based on issued GAO reports, reviews of agency documents related to the development of FPA, and discussions with agency officials.

Mr. Chairman and Members of the Subcommittee:

I am pleased to be here today to discuss key actions that we believe the federal wildland fire agencies—the Forest Service within the Department of Agriculture and four agencies within the Department of the Interior (Interior)—should take to improve their management of

* This Chapter.is an excerpted indexed edition of GAO Report GAO-08-433T, Dated February 12, 2008

wildland fires and help contain the rising costs of preparing for and responding to fires. Our nation's wildland fire problems have worsened in the past decade. Appropriations for wildland fire management activities tripled from about \$1 billion in fiscal year 1999 to more than \$3 billion in fiscal year 2007, while, on average, the acreage burned annually by wildland fires has increased by approximately 70 percent since the 1990s. As we have previously reported, a number of factors have contributed to worsening wildland fires and the increasing cost of managing fires, including an accumulation of fuels due in part to past fire suppression policies; severe weather and drought in some areas of the country, related in part to climate change; and growing numbers of homes built in or near wildlands, an area often called the wildland-urban interface. Concerns about the increasing cost of managing wildland fires, along with growing recognition of the long-term fiscal challenges facing the nation, have led Congress, GAO, and others to focus attention on ways to contain growing fire expenditures and to ensure that the agencies' wildland fire activities are appropriate and carried out in a cost-effective and efficient manner.

My testimony today summarizes findings from several of our recent reports, which identified key weaknesses in the agencies' management of wildland fires and recommended actions the agencies should take if they are to respond effectively. Specifically, I will focus on the agencies' efforts to (1) develop a cohesive strategy for preparing for and responding to wildland fires, (2) contain federal expenditures for preparing for and responding to wildland fires, and (3) set priorities and allocate funds for fuel reduction treatments. I will also discuss preliminary results from our ongoing work on the agencies' efforts to complete the interagency budget allocation and planning model known as fire program analysis (FPA), a key component of a cohesive strategy.

To evaluate these issues, we reviewed selected reports we have issued since 1999, and interviewed agency officials to identify recent actions the agencies have taken in the areas discussed in these reports. We also reviewed pertinent agency plans, policies, reports, and other documents, and interviewed federal and nonfederal officials, to obtain information on the development and status of FPA. The previous reports we reviewed, and our current performance audit—which was conducted from September 2007 to February 2008—were conducted in accordance with generally accepted government auditing standards. Those standards require that we plan and perform the audit to obtain sufficient, appropriate evidence to provide a reasonable basis for our findings and conclusions based on our audit objectives. We believe that the evidence obtained provides a reasonable basis for our findings and conclusions based on our audit objectives.

SUMMARY

In recent years, GAO has recommended a number of actions the federal wildland fire agencies should take to better understand the extent of, and address, the nation's wildland fire problems. These actions could also help contain the rising federal expenditures for responding to wildland fires. However, the agencies have not completed all needed actions. Specifically, we believe the agencies should

- ***recommit to developing a cohesive strategy that identifies options and associated funding to reduce fuels and address wildland fire problems.*** Despite our repeated calls for a cohesive wildland fire strategy, the agencies have not developed one—and do not intend to do so. In 1999, to address the problem of excess fuels and their potential to increase the severity of wildland fires and cost of suppression efforts, we recommended that a cohesive strategy be developed to identify the available long-term options for reducing fuels and the associated funding requirements.[1] By laying out various potential approaches for addressing wildland fire, the approximate costs associated with each approach, and the trade-offs involved among the various approaches, such a strategy would help Congress and the agencies make informed decisions about effective and affordable long-term approaches to addressing the nation's wildland fire problems. Six years later, in 2005, we reiterated the need for a cohesive strategy and broadened our recommendation's focus to better address the interrelated nature of fuel reduction efforts and wildland fire response.[2] However, although the agencies had consistently concurred with our recommendation to develop such a strategy, the Department of Agriculture's Under Secretary for Natural Resources and Environment testified in 2007 that he did not think completing these steps would be beneficial because it is too difficult to predict future vegetative conditions and agency priorities. Agency officials told us in January 2008 that they have no plans to develop the cohesive strategy we have called for.

- ***establish clear goals and a strategy to help contain wildland fire costs.*** In 2007, we reported that the Forest Service and Interior agencies were taking a number of steps intended to help contain wildland fire costs, including improving their decision support tools for helping officials select strategies for fighting wildland fires, but had not clearly defined their cost-containment goals or developed a strategy for achieving those goals—steps that are fundamental to sound program management.[3] In commenting on a draft of that report, the Forest Service and Interior identified several documents that they argue provide clearly defined goals and objectives that make up their strategy to contain costs.[4] However, these documents lack the clarity and specificity needed by officials in the field to help manage and contain wildland fire costs, and we therefore continue to believe that our recommendations, if effectively implemented, would help the agencies better manage their cost-containment efforts and improve their ability to contain wildland fire costs.
- ***continue to improve their processes for allocating fuel reduction funds and selecting fuel reduction projects.*** Also in 2007, we identified several shortcomings in the agencies' processes for allocating fuel reduction funds to field units and selecting fuel reduction projects, shortcomings that limit the agencies' ability to ensure that funds are directed where they will reduce risk most effectively.[5] The agencies have begun to take steps to improve the processes they use to allocate fuel reduction funds, but these efforts are incomplete and not fully coordinated. For example, several agencies are developing and implementing computer models to assist them in making allocation decisions, rather than relying primarily on historical funding patterns and professional judgment. Not all of the agencies have models, however, and none consistently uses models at national, regional, and local levels. The agencies also are taking steps to improve the information they use in allocating funds and selecting projects—including information on wildland fire risk and fuel treatment effectiveness—and to clarify the relative importance of the various factors they consider when making allocation decisions. We urge the agencies to continue

their efforts to improve their funding allocation and project selection processes so that they can more effectively use their limited fuel reduction dollars.

In addition, preliminary results from our ongoing work indicate that the FPA budget allocation model, which the agencies expect to complete in June 2008, may not allow the agencies to meet all of the key goals established for it. Specifically, our preliminary results suggest that FPA will not allow the agencies to analyze long-term trade-offs between annual fuel reduction treatments and the estimated costs of suppressing future large fires. GAO intends to conduct a full assessment of FPA once it is completed.

BACKGROUND

Although its effect on communities can be devastating, wildland fire is a natural and necessary process that provides many benefits to ecosystems, such as maintaining habitat diversity, recycling soil nutrients, limiting the spread of insects and disease, and promoting new growth by causing the seeds of fire-dependent species to germinate. Wildland fire also periodically removes brush, small trees, and other vegetation that can otherwise accumulate and increase the size, intensity, and duration of subsequent fires. Over the past century, however, various management practices—including fire suppression, grazing, and timber harvest—have reduced the normal frequency of fires in many forest and rangeland ecosystems and contributed to abnormally dense, continuous accumulations of vegetation, which can fuel uncharacteristically large or severe wildland fires. Federal researchers have estimated that unnaturally dense fuel accumulations on 90 million to 200 million acres of federal lands in the contiguous United States place these lands at an elevated risk of severe wildland fire.

In response to the growing wildland fire problem, the five federal agencies responsible for managing wildland fires—the Forest Service in the Department of Agriculture and the Bureau of Indian Affairs, Bureau of Land Management, Fish and Wildlife Service, and National Park Service in the Department of the Interior—adopted the 1995 federal wildland fire management policy, which formally recognized the essential role that fire plays in maintaining natural systems. This policy was subsequently reaffirmed and updated in 2001. Two important implications of the new policy are that the agencies recognized that (1) they needed to reduce accumulated vegetation that could fuel intense wildland fires and (2) it was not appropriate to continue attempting to suppress all fires.

Acknowledging the problem caused by accumulated fuels, Congress substantially increased appropriations for fuel reduction treatments—appropriating more than $3.2 billion to the Forest Service and Interior since 2001—and, in 2003, passed the Healthy Forests Restoration Act,[6] with the stated purpose of, among other things, reducing wildland fire risk to communities, municipal water supplies, and other at-risk federal land. After receiving its annual appropriation, the Forest Service allocates funds to its nine regional offices, which in turn allocate funds to individual national forests and grasslands. Interior, upon receiving its annual appropriation, allocates funds to its four fire management agencies—with the Bureau of Land Management receiving the largest share, about 50 percent of Interior's funding. Interior's agencies then allocate funds to their regional or state offices, which in turn allocate funds to individual field units, such as national parks or wildlife refuges. Forest Service and

Interior agency field units are generally responsible for selecting individual fuel reduction projects to undertake, which are typically conducted through mechanical treatments (using chainsaws, chippers, mowers, and the like) or by using prescribed fire (which land managers deliberately set to restore or maintain desired vegetative conditions). The agencies used the tools and fuel reduction funding provided by Congress to treat more than 18 million acres from 2001 through August 2007.

Over the last decade, Congress, the Office of Management and Budget, federal agency officials, and others have expressed concerns about mounting federal wildland fire expenditures. These concerns have led GAO, the Department of Agriculture's Office of Inspector General, the Forest Service, Interior, and others to conduct numerous reviews of the federal wildland fire program. These reviews identified many issues the agencies needed to address if they are to contain costs—issues generally related to reducing accumulated fuels, acquiring and using firefighting personnel and equipment, and selecting firefighting strategies.

Land managers and incident management teams (specialized fire-response teams that include personnel to handle command, planning, logistics, operations, and finance functions) have a wide spectrum of strategies available to them when responding to wildland fires, some of which can be significantly more costly than others. These strategies range from having a few personnel monitor a fire while allowing it to burn to achieve ecological benefits—a practice known as wildland fire use—to mobilizing all available personnel and equipment to try to control the entire perimeter of a fire or otherwise suppress it as quickly as possible. In selecting a strategy for a particular fire, land managers are required to consider the cost of suppression, the value of structures and other resources threatened by the fire, and the potential ecological effects of the fire. The agencies use the term "appropriate management response" for a strategy that considers these factors. Recent reports by GAO and others, however, have identified barriers to the agencies increasing their use of less aggressive strategies, which often cost less.

Agencies Need a Cohesive Wildland Fire Strategy to Manage Limited Resources Effectively, but Have Retreated from Their Commitment to Develop One

If the agencies and Congress are to make informed decisions about an effective and affordable long-term approach for addressing wildland fire, the agencies need a cohesive strategy that identifies the long-term options and associated funding for reducing excess vegetation and responding to fires. We first recommended the development of a cohesive strategy for addressing excess vegetation in 1999. By 2005, the agencies had yet to develop such a strategy, and that year we reiterated the need for such a strategy and broadened our recommendation's focus to include options not only for reducing fuels but also for responding to wildland fires when they do occur, in order to better address the interrelated nature of the two activities. We repeated our call for a cohesive strategy in 2006 and 2007.[7]

Although the agencies had consistently concurred with our recommendation to develop a cohesive strategy, in 2007 they retreated from their commitment to develop one. The Department of Agriculture's Under Secretary for Natural Resources and Environment testified before the Senate Committee on Energy and Natural Resources in January 2007, and before the House Subcommittee on National Parks, Forests and Public Lands in June 2007, that he did not think it useful to provide specific funding estimates for fuel treatments years into the future because conditions on the ground change over time and may change priorities in future years. Forest Service and Interior officials subsequently told us in January 2008 that they have no plans to develop a cohesive strategy that identifies long-term options and associated funding requirements.

Despite the agencies' retreat from their commitment to develop a cohesive strategy, a strategy of this sort nevertheless remains fundamental if the agencies and Congress are to fully understand the potential choices, and associated costs, for addressing wildland fire problems. We also believe the agencies have mischaracterized our recommendation to develop long-term options, and associated funding, for reducing fuels. Our intent was not to have the agencies identify the specific areas they would treat each year into the future, but rather that they develop broad options for reducing fuels, including estimated costs, and analyze the effects of the different options on the predicted costs of preparing for and responding to wildland fires in the future. One such analysis was developed in 2002 by a team of Forest Service and Interior experts, who produced an estimate of the funds needed to implement each of eight different fuel reduction options for protecting communities and ecosystems across the nation over the next century.[8] The team determined that effectively reducing the risks to communities and ecosystems across the nation could require an approximate tripling of fuel reduction funding, to about $1.4 billion annually, for an initial period of several years. These initially higher costs for fuel reduction would decline after fuels had been sufficiently reduced to allow less-expensive prescribed burning methods in many areas. More importantly, the team estimated that the reduction in fuels would allow the agencies to suppress more fires at lower cost and would reduce total wildland fire management costs and risk after 15 years. Alternately, the team concluded that maintaining the then-current level of investment in fuel reduction would increase costs as well as risks to communities and ecosystems in the long term. However, the Office of Management and Budget raised concerns about the accuracy of the long-term funding estimates used by the study; as a result, agency officials told us in 2006 that they needed to improve the data before they could develop a cohesive strategy. Now, however—and despite agency efforts to improve their data—this concern appears to be moot, as the agencies have abandoned their commitment to develop the strategy.

Lack of Clear Goals or a Strategy Hinders Federal Agencies' Efforts to Contain the Costs of Fighting Fires

We reported in 2007 that although the Forest Service and Interior agencies had taken several steps intended to help contain wildland fire costs, they had not clearly defined their cost-containment goals or developed a strategy for achieving those goals—steps that are fundamental to sound program management. As we reported, the agencies are implementing a number of steps designed to help them contain wildland fire costs—such as improving how they acquire and use firefighting assets, updating policies to require officials to consider the full spectrum of available strategies when selecting a firefighting strategy, and developing new decision support tools that help officials select the most appropriate strategy. However, we also found that the agencies had neither clearly defined the goals of their cost-containment efforts nor developed a clear plan for how the various steps they are taking to help contain costs fit together. Without such a strategy, we believe the agencies will have difficulty determining whether they are taking the most important steps first, as well as the extent to which the steps they are taking will help contain costs.

As a result, we recommended that the agencies take several steps to improve the management of their cost-containment efforts, including establishing clearly defined goals and measurable objectives and a strategy to achieve them. Because of the importance of these actions and continuing concerns about the agencies' response to the increasing cost of wildland fires—and so that the agencies could use the results of these actions to prepare for the 2008 fire season—we recommended the agencies provide Congress with this information no later than November 2007, a step they have yet to take. The Forest Service and Interior, in commenting on a draft of that report, generally disagreed with the characterization of many of our findings, but they neither agreed nor disagreed with our recommendations. In particular, they identified several agency documents that they argue provide clearly defined goals and objectives and that make up their strategy to contain costs. Although the documents cited by the agencies provide overarching goals and objectives, they lack the clarity and specificity needed by land management and firefighting officials in the field to help manage and contain wildland fire costs. Agency policy, for example, established an overarching goal of suppressing wildland fires at minimum cost, considering firefighter and public safety and the importance of resources being protected, but the agencies have established neither clear criteria by which to weigh the relative importance of the often-competing elements of this broad goal nor measurable objectives by which to determine if the agencies are meeting the goal. As a result, despite improvements the agencies continue to make to policy, decision support tools, and oversight, we believe that managers in the field lack a clear understanding of the relative importance that the agencies' leadership places on containing costs, and—as we concluded in our 2007 report—are therefore likely to continue to select firefighting strategies without due consideration of the costs of suppression. We continue to believe that our recommendations, if effectively implemented, would help the agencies better manage their cost-containment efforts and improve their ability to contain wildland fire costs.[9]

Better Information and a Systematic Process Would Improve Agencies' Approach to Allocating Fuel Reduction Funds

In 2007, we also identified several shortcomings in the agencies' processes for allocating fuel reduction funds to field units and selecting fuel reduction projects, which the agencies should correct in order to use their fuel reduction funds more effectively. Specifically, we noted that the agencies (1) did not consistently use systematic allocation processes—that is, processes that are based on criteria and applied consistently—in all agencies or at all levels, often relying instead on historical funding levels and professional judgment to allocate funds and select projects; (2) did not consistently consider the potential risk from wildland fire or the potential effectiveness of fuel reduction treatments when allocating funds and selecting projects; and (3) had not clarified the relative importance of the numerous factors they consider when allocating funds and selecting projects, including factors (such as funding stability or the use of forest products resulting from fuel reduction treatments) unrelated to risk or effectiveness.

Accordingly, we recommended that the agencies improve their allocation processes in three areas. First, we recommended that the agencies develop and routinely use a systematic allocation process that is based on criteria, applied consistently, and common to all the agencies. Second, we recommended that the agencies work to improve the information they use to make allocation decisions, particularly information on wildland fire risk and fuel treatment effectiveness. Third, we recommended that the agencies clarify the relative importance of the various factors they consider when allocating funds. Without improvements in these three areas, we noted that the agencies would likely continue relying on "allocation by tradition"—that is, allocating fuel reduction funds on the basis of past funding levels rather than on calculated need.

Some agencies have begun implementing systematic processes for allocating funds. In 2007, the Forest Service began using a computer model to influence funding allocations to its nine regions, and it continues to refine and expand its use of the model, including introducing improved data about the likelihood of fire in a particular area. In addition, all nine Forest Service regions are required, beginning in 2008, to use the model as part of their process for allocating funds to national forests. Interior is developing a similar computer model for allocating funds to its agencies, in part based on the Forest Service's model. For fiscal year 2007, Interior allocated 5 percent of its fuel reduction project funds to its four agencies using the model; for fiscal year 2008, according to an Interior official, Interior will use the model to allocate all of its fuel reduction project funds to its four agencies, within constraints designed to reduce the potential impact of funding changes.[10] Officials from both the Forest Service and Interior told us that the agencies are working closely with each other on model development. Of Interior's agencies, the Bureau of Land Management is developing a model similar to Interior's for allocating funds to its state offices; the Fish and Wildlife Service uses its own computer model when allocating funds to regional offices; the Bureau of Indian Affairs allocates funds to its regions using a formula that considers past performance and proposed work; and the National Park Service allocates funds to its regions primarily on the basis of historical funding levels. However, Interior is working to standardize the allocation process within these agencies as well; a department official told us that Interior plans to use its model to allocate funds down to the agencies' state and regional levels in fiscal year 2009.

Although the models some of the agencies are developing represent substantial steps forward in systematically allocating funds, these steps are incomplete and not fully coordinated. Specifically, not all the agencies have models; none consistently uses models at the national, regional, and local levels; and the models that are in use are not common to all agencies. Further, the models, even where used, often exert only a small influence on allocation decisions, partly because the agencies do not yet have full confidence in the models' data. Until the models serve as the foundation for allocation decisions, such decisions will continue to rely mainly on historical funding patterns and professional judgment. Accordingly, we urge the agencies to continue developing an allocation process that is systematic and that is common to all agencies.[11]

The agencies are also continuing to investigate ways to develop and use measures of risk and treatment effectiveness. Forest Service and Interior officials told us, for example, that researchers are looking at areas burned in past wildland fires to assess the extent to which fuel treatments altered fire behavior. Although efforts such as these are likely to be long-term undertakings and involve considerable research investment and activity, developing such measures would improve the agencies' ability to assess and compare the cost-effectiveness of potential treatments in deciding how to optimally allocate scarce funds. Finally, such information could also help the agencies address our third recommendation—that is, to clarify the relative importance of the various factors they consider when allocating funds. Such an effort is already under way at Interior, according to an Interior official, and the agency hopes to complete its work before the 2008 fire season. A Forest Service official stated that the Forest Service is also working to prioritize the various factors, but did not provide a timetable for completing this effort.

AGENCIES PLAN TO COMPLETE THE FPA MODEL IN 2008, BUT THE EXTENT TO WHICH IT WILL HELP ENSURE COST-EFFECTIVE FIRE MANAGEMENT DECISIONS IS UNCERTAIN

Agency officials plan to complete the FPA model by June 30, 2008, but preliminary results from our ongoing review raise questions about the extent to which the current model will be able to meet all of the key goals established for FPA. FPA—a common interagency performance-based system for program planning and budgeting for the full scope of fire management activities, including preparedness, large fire suppression, and fuel reduction treatments—was proposed and funded to address shortcomings that Congress, GAO, and the Office of Management and Budget identified in the agencies' existing budget allocation frameworks. FPA also is critical to developing and implementing a cohesive strategy,[12] and to the agencies' efforts to contain wildland fire costs. Development of FPA commenced in 2002. According to a 2001 report commissioned by the agencies that serves as the foundation of FPA,[13] FPA was intended to establish a common framework for the agencies to

- determine national budget needs by analyzing budget alternatives at the local level—using a common, interagency process for fire management planning and budgeting—and aggregating the results;
- determine the relative costs and benefits for the full scope of fire management activities, including potential trade-offs among investments in fuel reduction, fire preparedness, and fire suppression activities; and
- identify, for any given budget level, the most cost-effective mix of personnel and equipment to carry out these activities.

In addition, because responding to wildland fires often requires coordination and collaboration among federal, state, tribal, and local firefighting entities to effectively protect lives, homes, and resources, the agencies were directed to develop FPA in conjunction with their nonfederal partners and to recognize the availability of adjacent nonfederal firefighting resources when determining the appropriate amount and location of federal resources.

FPA program and senior agency officials told us that, when completed, FPA will allow the agencies to meet the key goals established for it, but preliminary results from our ongoing review have raised questions about FPA's ability to do so. In particular, FPA likely will analyze only 5 years of fuel reduction treatments when modeling the effect such treatments will have on future large fire events, according to FPA program officials, although they have not yet made a final determination on the number of years to be analyzed. The officials said that it is not possible to identify fuel treatment projects more than 5 years into the future with sufficient accuracy to include in the analysis. Such a limited time frame, however, substantially impairs the ability of the model to analyze long-term tradeoffs between annual fuel reduction treatment costs and future expected suppression costs for large fires, a key goal of FPA. Officials say that the FPA model expected to be completed in 2008 is the first step in an iterative development process and can be improved to increase its capability to analyze the trade-offs, but they could not provide a time frame for doing so.

In addition, in 2006, after 4 years of model development, the agencies initiated substantial changes to the process FPA will use to analyze needed firefighting resources and determine where best to locate these resources; they are also still deciding how senior officials will use the model's output to allocate funds between agencies and geographic regions of the country. It is not clear at this time the extent to which FPA will meet the key goal of identifying the most cost-effective allocation of resources for a given budget level, because the agencies are still developing the FPA model and determining how it will be used. A full assessment of FPA cannot be conducted, however, until the agencies complete the model; at that time, we plan on assessing the extent to which FPA will meet the key goals established.

CONCLUSIONS

Faced with an incendiary mix of accumulated fuels, climate change, and burgeoning development in fire-prone areas, and constrained by our nation's long-term fiscal outlook, the federal wildland fire agencies need to commit to a more considered, long-term approach to managing their resources in order to address the wildland fire problem more effectively and efficiently. They have taken an important first step by establishing and updating federal wildland fire policy. Development of strategies and management tools for agency officials to use in achieving the policy's vision, however, has been uneven. The agencies are making progress in certain areas, including improving funding allocation processes for reducing fuels and requiring appropriate management response to fires that occur. In addition, the agencies are continuing to develop FPA, which, if implemented appropriately, could significantly improve the agencies' ability to allocate their resources effectively. But broader efforts have stalled—as in the development of cost containment goals and objectives—or even lost ground, as evidenced by the agencies' retreat from their earlier commitment to develop the cohesive wildland fire strategy we have called for. If the agencies are to achieve lasting results in their efforts to address the wildland fire problem, they will need a sustained commitment by agency leadership to developing both a long-term strategy that identifies potential options (and their costs) for managing wildland fires and the tools for carrying out such a strategy.

Mr. Chairman, this concludes my prepared statement. I would be pleased to answer any questions that you or other Members of the Subcommittee may have at this time.

REFERENCES

[1] GAO, Western National Forests: A Cohesive Strategy Is Needed to Address Catastrophic Wildfire Threats, GAO/RCED-99-65 (Washington, D.C.: Apr. 2, 1999).

[2] GAO, Wildland Fire Management: Important Progress Has Been Made, but Challenges Remain to Completing a Cohesive Strategy, GAO-05-147 (Washington, D.C.: Jan. 14, 2005).

[3] GAO, Wildland Fire Management: Lack of Clear Goals or a Strategy Hinders Federal Agencies' Efforts to Contain the Costs of Fighting Fires, GAO-07-655 (Washington, D.C.: June 1, 2007).

[4] Department of the Interior, Department of Agriculture, Department of Energy, Department of Defense, Department of Commerce, Environmental Protection Agency, Federal Emergency Management Agency, and National Association of State Foresters, Review and Update of the 1995 Federal Wildland Fire Management Policy (Washington, D.C.: January 2001). Department of Agriculture, Department of the Interior, and Western Governors' Association, A Collaborative Approach for Reducing Wildland Fire Risks to Communities and the Environment, 10-Year Strategy Implementation Plan (Washington, D.C.: December 2006).

[5] GAO, Wildland Fire Management: Better Information and a Systematic Process Could Improve Agencies' Approach to Allocating Fuel Reduction Funds and Selecting Projects, GAO-07-1168 (Washington, D.C.: Sept. 28, 2007).

[6] Pub. L. No. 108-148 (2003).

[7] GAO, Wildland Fire Management: Update on Federal Agency Efforts to Develop a Cohesive Strategy to Address Wildland Fire Threats, GAO-06-671R (Washington, D.C.: May 1, 2006); and Wildland Fire Management: Lack of Cohesive Strategy Hinders Agencies' Cost Containment Efforts, GAO-07-427T (Washington, D.C.: Jan. 30, 2007).

[8] Wendell Hann, et al, A Cohesive Strategy for Protecting People and Sustaining Natural Resources: Predicting Outcomes for Program Options (a paper presented at the Fire, Fuel Treatments, and Ecological Restoration Conference, a meeting on national wildland fire experts convened by the Forest Service's Rocky Mountain Research Station, Fort Collins, Colorado, April 2002).

[9] State and local governments also play an important role in containing fire suppression costs. As we have noted, protecting the increasing number of homes in the wildland urban interface at risk from wildland fire has contributed significantly to the increased cost of managing wildland fires. See GAO, Wildland Fire Suppression: Lack of Clear Guidance Raises Concerns about Cost Sharing between Federal and Nonfederal Entities, GAO-06-570 (Washington, D.C.: May 30, 2006). We reported that federal officials were concerned, and some nonfederal officials acknowledged, that the framework federal and nonfederal agencies use to share the costs of suppressing wildland fires affecting both federal and nonfederal land—combined with the availability of federal emergency funding to reimburse nonfederal agencies for the cost of fighting such fires—insulates state and local governments from the cost of providing fire protection. A bill introduced in the Senate in November 2007, S. 2390, would provide an incentive to local governments by allowing for greater federal reimbursement of firefighting costs in

communities that adopt a model fire ordinance, including requirements for fire-safe building materials and maintenance of defensible space around homes.

[10] In order to minimize budget-related disruptions resulting from significant funding changes, Interior will limit such changes to no more than 15 percent of each agency's budget.

[11] As we noted in our 2007 report, although we believe the agencies should use a common allocation process, the process may need to be customized to reflect differences among the agencies, including differences in scale, variety of ecosystems managed, agency mission, and other factors.

[12] Other elements critical to developing the cohesive strategy include LANDFIRE—a geospatial data and modeling system designed to generate comprehensive maps of vegetation, fire, and fuel characteristics nationally—and fire management plans, which are local plans developed by individual agency management units (such as wildlife refuges or national forests) to define each unit's program to prepare for and respond to wildland fires.

[13] U.S. Department of Agriculture, Forest Service, and U.S. Department of the Interior, Developing an Interagency, Landscape-Scale Fire Planning Analysis and Budget Tool (Washington, D.C., November 2001).

Chapter 4

WILDLAND FIRE MANAGEMENT: IMPROVEMENTS COULD ENHANCE FEDERAL AGENCIES' EFFORTS TO CONTAIN THE COSTS OF FIGHTING FIRES *

WHY GAO DID THIS STUDY

Annual appropriations to prepare for and respond to wildland fires have increased substantially over the past decade, in recent years totaling about $3 billion. The Forest Service within the Department of Agriculture and four agencies within the Department of the Interior (Interior) are responsible for responding to wildland fires on federal lands. GAO determined what steps federal agencies have taken to (1) address key operational areas that could help contain the costs of preparing for and responding to wildland fires and (2) improve their management of their cost-containment efforts. This testimony is based on GAO's June 2007 report, *Wildland Fire Management: Lack of Clear Goals or a Strategy Hinders Federal Agencies' Efforts to Contain the Costs of Fighting Fires* (GAO-07- 655).

Mr. Chairman and Members of the Committee:

I am pleased to be here today to discuss federal firefighting agencies' efforts to contain the costs of preparing for and responding to wildland fires—costs that have increased substantially over the past decade. Wildland fire appropriations to prepare for and respond to wildland fires, including appropriations for reducing fuels, have increased from an average of $1.1 billion annually from fiscal years 1996 through 2000 to an average of more than $2.9 billion annually from fiscal years 2001 through 2005; adjusted for inflation, these appropriations increased from $1.3 billion to $3.1 billion.[1] Accumulations of fuels, due in part to past suppression policies; severe drought and weather in some areas of the country; and continued development in or near wildlands—an area commonly known as the wildland-urban interface—have contributed to increased costs. Five federal land management

* This Chapter is an excerpted indexed edition of GAO Report GAO-07-922T, Dated June 25, 2007

agencies—the Forest Service within the Department of Agriculture (Agriculture) and the Bureau of Land Management, Bureau of Indian Affairs, National Park Service, and Fish and Wildlife Service within the Department of the Interior (Interior)—are responsible for managing wildland fires on federal lands. Congress, the Office of Management and Budget, federal agency officials, and others have expressed concerns about the mounting federal wildland fire expenditures. These concerns have led federal agencies (including the Forest Service, Interior, the Agriculture Office of Inspector General, and GAO) and others to conduct numerous reviews of the federal wildland fire program.

My testimony is based on our report, released today, that discusses steps the Forest Service and Interior agencies have taken to (1) address key operational areas that could help contain the costs of preparing for and responding to wildland fires and (2) improve their management of their cost-containment efforts.[2] I presented the preliminary results of our work before this Committee in January 2007.[3]

SUMMARY

In summary, the Forest Service and Interior agencies have initiated a number of steps to address key operational areas that past studies identified as needing improvement to help federal agencies contain wildland fire costs, but the effects on containing costs are unknown, in part because many of these steps are not yet complete. For example,

- Federal firefighting agencies are developing a system to help them better identify and set priorities for lands needing treatment to reduce fuels. The agencies are developing, but have not yet finalized, a plan for keeping data in the system current.
- Federal agencies have also taken some steps to improve how they acquire and use personnel, equipment, and other firefighting assets, such as implementing a computerized system to more efficiently dispatch and track available firefighting assets. The agencies, however, have not completed the more fundamental step of determining the appropriate type and quantity of firefighting assets needed for the fire season. Over the past several years, the agencies have been developing a system for doing so, although we have concerns that recent modifications to the system may not allow the agencies to fully meet certain key goals.
- The agencies have clarified certain policies and are improving analytical tools that assist officials in identifying and implementing an appropriate response to a given fire. Other policies, however, limit the agencies' use of less aggressive firefighting strategies, which typically cost less.
- Federal agencies, working with nonfederal entities, have recently taken steps to clarify guidance to better ensure that firefighting costs are shared consistently for fires that threaten both federal and nonfederal lands and resources, although it is unclear how the agencies will provide oversight to ensure that this guidance is followed in the field.

Despite steps taken to strengthen the management of their cost-containment efforts, the agencies have neither clearly defined their cost-containment goals and objectives nor developed a strategy for achieving them—steps that are fundamental to sound program management. Although the agencies have established a broad goal of suppressing wildland fires at minimum cost—considering firefighter and public safety, and resources and structures to be protected—they have no defined criteria by which to weigh the relative importance of these often-competing priorities. As a result, according to agency officials and reports, officials in the field lack a clear understanding of the relative importance the agencies' leadership places on containing costs and, therefore, are likely to select firefighting strategies without due consideration of the costs of suppression. The agencies have also yet to develop a vision of how the various cost-containment steps they are taking relate to one another or to determine the extent to which these steps will be effective. The agencies are working to develop a better cost-containment performance measure, but the measure may take a number of years to fully refine. Finally, the agencies have taken, or are beginning to take, steps to improve their oversight and increase accountability—such as requiring agency officials to evaluate firefighting teams on how well they contain costs—although the extent to which these steps will assist the agencies in containing costs is unknown.

We recommended in our report that the Secretaries of Agriculture and the Interior take several steps to improve their management of their cost-containment efforts. The Forest Service and Interior generally disagreed with our findings, stating that we had not accurately portrayed some of the agencies' actions to contain costs; they neither agreed nor disagreed with our recommendations. We continue to believe that our recommendations, if effectively implemented, would help the agencies better manage their cost-containment efforts and improve their ability to contain wildland fire costs.

BACKGROUND

Over the past decade, the number of acres burned annually by wildland fires in the United States has substantially increased. Federal appropriations to prepare for and respond to wildland fires, including appropriations for fuel treatments, have almost tripled. Increases in the size and severity of wildland fires, and in the cost of preparing for and responding to them, have led federal agencies to fundamentally reexamine their approach to wildland fire management. For decades, federal agencies aggressively suppressed wildland fires and were generally successful in decreasing the number of acres burned. In some parts of the country, however, rather than eliminating severe wildland fires, decades of suppression contributed to the disruption of ecological cycles and began to change the structure and composition of forests and rangelands, thereby making lands more susceptible to fire.

Increasingly, the agencies have recognized the role that fire plays in many ecosystems and the role that it could play in the agencies' management of forests and watersheds. The agencies worked together to develop a federal wildland fire management policy in 1995, which for the first time formally recognized the essential role of fire in sustaining natural systems; this policy was subsequently reaffirmed and updated in 2001. The agencies, in conjunction with Congress, also began developing the National Fire Plan in 2000.[4] To align their policies and to ensure a consistent and coordinated effort to implement the federal wildland fire policy and National Fire Plan, Agriculture and Interior established the Wildland Fire Leadership Council in 2002.[5] In addition to noting the negative effects of past successes in suppressing wildland fires, the policy and plan also recognized that continued development in the wildland-urban interface has placed more structures at risk from wildland fire at the same time that it has increased the complexity and cost of wildland fire suppression. Forest

Service and university researchers estimated in 2005 that about 44 million homes in the lower 48 states are located in the wildland-urban interface.

To help address these trends, current federal policy directs agencies to consider land management objectives—identified in land and fire management plans developed by each local unit, such as a national forest or a Bureau of Land Management district—and the structures and resources at risk when determining whether or how to suppress a wildland fire. When a fire starts, the land manager at the affected local unit is responsible for determining the strategy that will be used to respond to the fire. A wide spectrum of strategies is available to choose from, some of which can be significantly more costly than others. For example, the agencies may fight fires ignited close to communities or other high-value areas more aggressively than fires on remote lands or at sites where fire may provide ecological or fuel-reduction benefits. In some cases, the agencies may simply monitor a fire, or take only limited suppression actions, to ensure that the fire continues to pose little threat to important resources, a practice known as "wildland fire use."

Federal Agencies Are Taking Some Steps to Contain Wildland Fire Costs, but Results Are Unknown

The Forest Service and Interior agencies have initiated a number of steps to address issues that we and others have identified as needing improvement to help federal agencies contain wildland fire costs, but the effects of these steps on containing costs are unknown, in part because many of the steps are not yet complete. Dozens of studies by federal agencies and other organizations examining federal agencies' management of wildland fire have repeatedly identified a number of similar issues needing improvement to help contain wildland fire costs. These issues generally fall into one of three operational areas—reducing accumulated fuels, acquiring and using firefighting assets, and selecting firefighting strategies. Recent studies have also raised concerns about the framework used to share the cost of fighting fires between federal and nonfederal entities.

First, federal firefighting agencies have made progress in developing a system to help them better identify and set priorities for lands needing treatment to reduce accumulated fuels. Many past studies have identified fuel reduction as important for containing wildland fire costs because accumulated fuels can contribute to more-severe and more costly fires. The agencies are developing a geospatial data and modeling system, called LANDFIRE, intended to produce consistent and comprehensive maps and data describing vegetation, wildland fuels, and fire regimes across the United States.[6] The agencies will be able to use this information to help identify fuel accumulations and fire hazards across the nation, help set nationwide priorities for fuel-reduction projects, and assist in determining an appropriate response when wildland fires do occur. According to Forest Service and Interior officials, the agencies completed mapping the western United States in April 2007; mapping of the eastern states is scheduled to be completed by 2008 and of Alaska and Hawaii by 2009. The agencies, however, have not yet finalized their plan for ensuring that collected data are routinely updated to reflect changes to fuels, including those from landscape-altering events, such as hurricanes, disease, or wildland fires themselves. Forest Service and Interior officials told us that they recognize the importance of ensuring that data are periodically updated and are developing a plan to operate and maintain the system, including determining how often data will be updated. The agencies expect to submit this plan to the Wildland Fire Leadership Council for approval in June 2007.

Second, the agencies have also taken some steps to improve how they acquire and use firefighting personnel, aviation resources, and equipment—assets that constitute a major cost of responding to wildland fires—but much remains to be done. The agencies have improved their systems for dispatching and monitoring firefighting assets and for gathering and analyzing cost data. However, they have yet to complete the more fundamental step of determining the appropriate type and quantity of firefighting assets needed for the fire season. Over the past several years, the agencies have been developing a Fire Program Analysis (FPA) system, which was proposed and funded to help the agencies

- determine national budget needs by analyzing budget alternatives at the local level—using a common, interagency process for fire management planning and budgeting—and aggregating the results;
- determine the relative costs and benefits for the full scope of fire management activities, including potential trade-offs among investments in fuel reduction, fire preparedness, and fire suppression activities; and
- identify, for a given budget level, the most cost-effective mix of personnel and equipment to carry out these activities.

We have said for several years—and the agencies have concurred—that FPA is critical to helping the agencies contain wildland fire costs and plan and budget effectively. Recent design modifications to the system, however, raise questions about the agencies' ability to fully achieve these key goals. A midcourse review of the developing system resulted in the Wildland Fire Leadership Council's approving in December 2006 modifications to the system's design. FPA and senior Forest Service and Interior officials told us in April 2007 they believed the modifications will allow the agencies to meet the key goals. The officials said they expected to have a prototype developed for the council's review in June 2007 and to substantially complete the system by June 2008. We have yet to systematically review the modifications, but after reviewing agency reports on the modifications and interviewing knowledgeable officials, we have concerns that the modifications may not allow the agencies to meet FPA's key goals. For example, under the redesigned system, local land managers will use a different method to analyze and select various budget alternatives, and it is unclear whether this method will identify the most cost-effective allocation of resources. In addition, it is unclear how the budget alternatives for local units will be meaningfully aggregated on a nationwide basis, a key FPA goal.

Third, the agencies have clarified certain policies and are improving analytical tools to assist agency officials in identifying and implementing an appropriate response to a given fire. Officials have a wide spectrum of strategies available to them when responding to wildland fires, some of which can be significantly more costly than others. For individual fires, past studies have found that officials may not always consider the full range of available strategies and may not select the most appropriate one, which would consider the cost of suppression; value of structures and other resources threatened by the fire; and, where appropriate, any benefits the fire may provide to natural resources. The agencies call a strategy that considers these factors the "appropriate management response." The agencies updated their policies in 2004 to require officials to consider the full spectrum of available strategies when selecting one to use. Nevertheless, other policies limit the agencies' use of less aggressive strategies, which typically cost less. The Forest Service and Interior agencies are working together to revise these policies—revisions that could, for example, allow different areas of the same fire to be managed for suppression and wildland fire use concurrently or allow a fire that was previously being suppressed to be managed instead for wildland fire use. The agencies are also continuing to refine existing tools, and to develop new ones, for analyzing both fuel and predicted weather conditions to model expected fire behavior, information that officials can use to identify appropriate suppression strategies; these tools are still being designed and tested. It is still too early to tell, however, to what extent the policy changes being considered or the new tools being developed will help to contain costs.

Finally, we and others have also reported that the existing framework for sharing firefighting costs between federal and nonfederal entities insulates state and local governments from the cost of protecting homes and communities in or near wildlands, which may reduce those governments' incentive to adopt building codes and land use requirements that could help reduce the cost of suppressing wildland fires.[7] Federal agencies, working with nonfederal entities, have recently taken steps to clarify guidance and better ensure that firefighting costs are shared consistently for fires that threaten both federal and nonfederal lands and resources. In early 2007, the Forest Service and Interior agencies approved an updated template that land managers can use when developing master agreements—which establish the framework for sharing costs between federal and nonfederal entities—as well as agreements on how to share costs for a specific fire. Because master agreements are normally updated every 5 years, however, it may take several years to fully incorporate this new guidance. Although the new guidance states that managers must document their rationale for selecting a particular cost-sharing method, officials told us that the agencies have no clear plan for how they will provide oversight to ensure that appropriate cost-sharing methods are used.

Lack of Clear Goals or a Strategy Hinders Federal Agencies' Management of Wildland Fire Cost-Containment Efforts

Despite steps taken to strengthen their management of cost-containment efforts, the agencies have neither clearly defined their cost-containment goals and objectives nor developed a strategy for achieving them—steps that are fundamental to sound program management. To manage their cost-containment efforts effectively, the Forest Service and Interior agencies should, at a minimum, have (1) clearly defined goals and measurable objectives, (2) a strategy to achieve the goals and objectives, (3) performance measures to track their progress, and (4) a framework for holding appropriate agency officials accountable for achieving the goals.[8]

First, although the agencies have established a broad goal of suppressing wildland fires at minimum cost considering firefighter and public safety and the resources and structures to be protected, they have established neither clear criteria by which to weigh the relative importance of these often-competing priorities nor measurable objectives by which to determine if they are meeting their goal. Without such criteria and objectives, according to agency officials we interviewed and reports we reviewed, officials in the field lack a clear understanding of the relative importance that the agencies' leadership places on containing costs and, therefore, are likely to select firefighting strategies without due consideration of costs.

Second, the agencies have yet to establish an overall cost-containment strategy. Without a strategy designed to achieve clear cost-containment goals, the agencies (1) have no assurance that the variety of steps they are taking to help contain wildland fire costs are prioritized so that the most important steps are undertaken first and (2) are unable to determine to what extent these steps will help contain costs and if a different approach may therefore be needed.

Third, the agencies recently adopted a new performance measure—known as the stratified cost index—that may improve the agencies' ability to evaluate their progress in containing costs, but the measure may take a number of years to fully refine. Also, although the agencies have in recent years improved their data on suppression costs and fire characteristics, additional improvement is needed. In particular, cost data for "fire complexes"—that is, two or more fires burning in proximity that are managed as a single incident—are particularly difficult to identify. Thus, the costs of many of the largest fires are not included in the index, limiting its effectiveness. Further, to date, the index is based solely on fires managed by the Forest Service. Forest Service researchers are currently developing, at Interior's request, a similar index for fires managed by the Interior agencies, but it will be several years, at the earliest, before enough data have been collected for the index to be useful. In addition, because the stratified cost index is based on costs from previous fires—and because the agencies have only recently begun to emphasize the importance of using less aggressive suppression strategies—we are concerned that the index does not include data from many fires where less costly firefighting strategies were used. As a result, the index may not accurately identify fires where more, or more-expensive, resources were used than needed. According to Forest Service officials, data from recent fires will be added annually; over time, the index should therefore include more fires where less aggressive firefighting strategies were used.

Finally, the agencies have also taken, or are beginning to take, steps to improve their oversight and accountability framework, although the extent to which these steps will assist the agencies in containing costs is unknown. For example, the agencies have issued guidance clarifying that land managers, not fire managers, have primary responsibility for containing wildland fire costs, but they have not yet determined how the land managers are to be held accountable for doing so. Rather, the agencies have taken several incremental steps intended to assist land managers in carrying out this responsibility—such as assigning "incident business advisors" to observe firefighting operations and work with fire managers to identify ways those operations could be more cost-effective, and requiring land managers to evaluate fire managers for how well they achieve cost-containment goals. The utility of these steps, however, may be limited because the agencies have yet to establish a clear measure to evaluate the benefits and costs of alternative firefighting strategies. Some past studies have concluded that the absence of such a measure fundamentally weakens the agencies' ability to provide effective oversight.

CONCLUSIONS

Continuing concerns about the cost of preparing for and responding to wildland fires have spurred numerous studies and actions by federal wildland fire agencies, but little in the way of a coordinated and focused effort to rein in these costs. Although the agencies have taken—and continue to take—steps intended to contain wildland fire costs, the effect of these steps on containing costs is unknown, in part because the agencies lack a clear vision for what they want to achieve. Without clearly defined cost-containment goals and objectives, federal land and fire managers in the field are more likely to select strategies and tactics that favor suppressing fires quickly over those that seek to balance the benefits of protecting the resources at risk and the costs of protecting them. Further, without clear goals, the agencies will be unable to develop consistent standards by which to measure their performance. Perhaps most important, without a clear vision of what they are trying to achieve and a systematic approach for achieving it, the agencies—and Congress and the American people—have little assurance that cost-containment efforts will lead to substantial improvement.

Thus, to help the agencies manage their ongoing efforts to contain wildland fire costs effectively and efficiently, and to assist Congress in its oversight role, we recommended in our report that the Secretaries of Agriculture and the Interior work together to direct their respective agencies to (1) establish clearly defined goals and measurable objectives for containing wildland fire costs, (2) develop a strategy to achieve these goals and objectives, (3) establish performance measures that are aligned with these goals and objectives, and (4) establish a framework to ensure that officials are held accountable for achieving the goals and objectives. Because of the importance of these actions and continuing concerns about the agencies' response to the increasing cost of wildland fires—and so that the agencies can use the results of these actions to prepare for the 2008 fire season—the agencies should provide Congress with this information no later than November 2007.

In commenting on a draft of our report, the Forest Service and Interior generally disagreed with the characterization of many of our findings; they neither agreed nor disagreed with our recommendations. In particular, the Forest Service and Interior stated that they did not believe we had accurately portrayed some of the significant actions they had taken to contain wildland fire costs, and they identified several agency documents that they believe provide clearly defined goals and objectives that make up their strategy to contain costs. Although documents cited by the agencies provide overarching goals and objectives, we believe that they lack the clarity and specificity needed by their land management and firefighting officials in the field to help manage and contain wildland fire costs. Therefore, we believe that our recommendations, if effectively implemented, would help the agencies better manage their cost-containment efforts and improve their ability to contain wildland fire costs.

Mr. Chairman, this concludes my prepared statement. I would be please to answer any questions that you or other Members of the Committee may have at this time.

REFERENCES

[1] Federal expenditures are a more direct measure of the federal government's investment in wildland fire activities, but the Forest Service and Interior agencies were unable to provide us with consistent data on these expenditures for the years we reviewed. As a

result, we are instead reporting appropriations data. We adjusted the appropriations dollars for inflation, using the chain-weighted gross domestic product price index with fiscal year 2005 as the base year.

[2] GAO, *Wildland Fire Management: Lack of Clear Goals and a Strategy Hinders Federal Agencies' Efforts to Contain the Costs of Fighting Fires*, GAO-07-655 (Washington, D.C.: June 1, 2007).

[3] GAO, *Wildland Fire Management: Lack of a Cohesive Strategy Hinders Agencies' Cost-Containment Efforts*, GAO-07-427T (Washington, D.C.: Jan. 30, 2007).

[4] The National Fire Plan is a joint interagency effort to respond to wildland fires. Its core comprises several strategic documents, including (1) a September 2000 report from the Secretaries of Agriculture and the Interior to the President in response to the wildland fires of 2000, (2) congressional direction accompanying substantial new appropriations in fiscal year 2001, and (3) several approved and draft strategies to implement all or parts of the plan.

[5] The Wildland Fire Leadership Council is composed of senior Agriculture and Interior officials, including the Agriculture Undersecretary for Natural Resources and Environment; the Interior Assistant Secretary for Policy, Management, and Budget; the Interior Deputy Assistant Secretary for Business Management and Wildland Fire; and the heads of the five federal firefighting agencies. Other members include representatives of the Intertribal Timber Council, the National Association of State Foresters, and the Western Governors' Association.

[6] A fire regime generally classifies the role that wildland fire plays in a particular ecosystem on the basis of certain characteristics, such as the average number of years between fires and the typical severity of fire under historic conditions.

[7] GAO, *Wildland Fire Suppression: Lack of Clear Guidance Raises Concerns about Cost Sharing between Federal and Nonfederal Entities*, GAO-06-570 (Washington, D.C.: May 30, 2006).

[8] Principles of sound program management for federal agencies are established in, among other sources, the Government Performance and Results Act of 1993 and GAO, *Standards for Internal Control in the Federal Government,* GAO/AIMD-00-21.3.1 (Washington, D.C.: November 1999).

Chapter 5

WILDLAND FIRE MANAGEMENT: A COHESIVE STRATEGY AND CLEAR COST-CONTAINMENT GOALS ARE NEEDED FOR FEDERAL AGENCIES TO MANAGE WILDLAND FIRE ACTIVITIES EFFECTIVELY *

WHY GAO DID THIS STUDY

Increasing wildland fire threats to communities and ecosystems, combined with rising costs of addressing those threats—trends that GAO and others have reported on for many years—have not abated. On average, the acreage burned annually by wildland fires from 2000 to 2005 was 70 percent greater than the acreage burned annually during the 1990s. Annual appropriations to prepare for and respond to wildland fires have also increased substantially over the past decade, totaling about $3 billion in recent years. The Forest Service within the Department of Agriculture and four agencies within the Department of the Interior (Interior) are responsible for responding to wildland fires on federal lands.

This testimony summarizes several key actions that federal agencies need to complete or take to strengthen their management of wildland fire program, including the need to (1) develop a long-term, cohesive strategy to reduce fuels and address wildland fire problems and (2) improve the management of their efforts to contain the costs of preparing for and responding to wildland fires. The testimony is based on several previous GAO reports and testimonies addressing wildland fire issues.

Mr. Chairman and Members of the Subcommittee:

I am pleased to be here today to discuss the key steps that we believe federal wildland fire agencies—the Forest Service within the Department of Agriculture and four agencies[1] within the Department of the Interior (Interior)—need to complete to manage their efforts to prepare for and respond to wildland fires effectively. Increasing wildland fire threats to

* This Chapter is an excerpted indexed edition of GAO Report GAO-07-1017T, Dated June 19, 2007

communities and ecosystems, combined with rising costs of addressing those threats—trends that we and others have reported on for many years—have not abated. On average, the acreage burned annually by wildland fires from 2000 to 2005 was 70 percent greater than the acreage burned annually during the 1990s. Appropriations to federal agencies to prepare for and respond to wildland fires, including appropriations for reducing fuels, have also increased substantially, from an average of $1.1 billion annually from fiscal years 1996 through 2000 to an average of more than $2.9 billion annually from fiscal years 2001 through 2005 (adjusted for inflation, these appropriations increased from $1.3 billion to $3.1 billion). A number of factors have contributed to more-severe fires and corresponding increases in expenditures for wildland fire management activities. These factors include an accumulation of fuels due to past fire suppression policies; severe weather and drought in some areas of the country; and growing numbers of homes built in or near wildlands, an area known as the wildland-urban interface. In light of the federal deficit and the long-term fiscal challenges facing the nation, attention has increasingly focused on ways to contain these growing expenditures and to ensure that federal agencies' wildland fire activities are appropriate and carried out in a cost-effective and efficient manner.

My testimony today is based on several of our previous reports and testimonies, which together discuss key issues we have identified over the last 7 years in federal agencies' management of wildland fire and critical actions the agencies need to complete if they are to effectively manage their efforts to prepare for and respond to wildland fires.[2] Specifically, my testimony focuses on the Forest Service and Interior agencies' (1) efforts to develop a long-term, cohesive strategy to reduce fuels and address wildland fire problems and (2) management of their efforts to contain the costs of preparing for and responding to wildland fires.

SUMMARY

In summary, the Forest Service and Interior agencies need to complete several actions to strengthen their overall management of the wildland fire program. Because a substantial investment and decades of work will be required to address wildland fire problems that have been decades in the making, the agencies need to develop a cohesive strategy that addresses the full range of wildland fire management activities. Such a strategy should identify the available long-term options and associated funding for reducing excess vegetation and responding to wildland fires if the agencies and the Congress are to make informed decisions about an effective and affordable long-term approach for addressing problems that have been decades in the making. We first recommended in 1999 that such a strategy be developed to address the problem of excess fuels and their potential to increase the severity of wildland fires and cost of suppression efforts.[3] By 2005, the agencies had yet to develop such a strategy, and we reiterated the need for a cohesive strategy and broadened our recommendation's focus to better address the interrelated nature of fuel reduction efforts and wildland fire response. Further, because the agencies said they would be unable to develop a cohesive strategy until they have completed certain key tasks, we recommended that the agencies develop a tactical plan outlining these tasks and the time frames needed for completing each task and a cohesive strategy.[4] Although the agencies concurred with our recommendations, as of April 2007, a tactical plan had yet to be developed.

Second, as we testified before the Senate Committee on Energy and Natural Resources in January 2007, the steps the Forest Service and Interior agencies have taken to date to contain wildland fire costs lack several key elements fundamental to sound program management, such as clearly defining cost-containment goals, developing a strategy for achieving those goals, and measuring progress toward achieving them. For cost-containment efforts to be effective, the agencies need to integrate cost-containment goals with the other goals of the wildland fire program—such as protecting life, resources, and property—and to recognize that trade-offs will be needed to meet desired goals within the context of fiscal constraints. Further, because cost-containment goals need to be considered in relation to other wildland fire program goals, it is important that the agencies integrate cost-containment goals within an overall cohesive strategy. Our forthcoming report on federal agencies' efforts to contain wildland fire costs includes more-detailed findings and recommendations to the agencies to improve the management of their cost-containment efforts; this report is expected to be released at a hearing before the Senate Committee on Energy and Natural Resources scheduled for June 26, 2007.

Background

Over the past decade, the number of acres burned annually by wildland fires in the United States has substantially increased. Federal appropriations to prepare for and respond to wildland fires, including appropriations for fuel treatments, have almost tripled. Increases in the size and severity of wildland fires, and in the cost of preparing for and responding to them, have led federal agencies to fundamentally reexamine their approach to wildland fire management. For decades, federal agencies aggressively suppressed wildland fires and were generally successful in decreasing the number of acres burned. In some parts of the country, however, rather than eliminating severe wildland fires, decades of suppression contributed to the disruption of ecological cycles and began to change the structure and composition of forests and rangelands, thereby making lands more susceptible to fire.

Increasingly, federal agencies have recognized the role that fire plays in many ecosystems and the role that it could play in the agencies' management of forests and watersheds. The agencies worked together to develop a federal wildland fire management policy in 1995, which for the first time formally recognized the essential role of fire in sustaining natural systems; this policy was subsequently reaffirmed and updated in 2001. The agencies, in conjunction with Congress, also began developing the National Fire Plan in 2000.[5] To align their policies and to ensure a consistent and coordinated effort to implement the federal wildland fire policy and National Fire Plan, Agriculture and Interior established the Wildland Fire Leadership Council in 2002.[6] In addition to noting the negative effects of past successes in suppressing wildland fires, the policy and plan also recognized that continued development in the wildland-urban interface has placed more structures at risk from wildland fire at the same time that it has increased the complexity and cost of wildland fire suppression. Forest Service and university researchers estimated in 2005 that about 44 million homes in the lower 48 states are located in the wildland-urban interface.

To help address these trends, current federal policy directs agencies to consider land management objectives—identified in land and fire management plans developed by each local unit, such as a national forest or a Bureau of Land Management district—and the structures and resources at risk when determining whether or how to suppress a wildland fire. When a fire starts, the land manager at the affected local unit is responsible for determining the strategy that will be used to respond to the fire. A wide spectrum of strategies is available, some of which can be significantly more costly than others. For example, the agencies may fight fires ignited close to communities or other high-value areas more aggressively than fires on remote lands or at sites where fire may provide ecological or fuel-reduction benefits. In some cases, the agencies may simply monitor a fire, or take only limited suppression actions, to ensure that the fire continues to pose little threat to important resources, a practice known as "wildland fire use."

AGENCIES NEED A COHESIVE STRATEGY TO ADDRESS WILDLAND FIRE PROBLEMS

Federal firefighting agencies need a cohesive strategy for reducing fuels and addressing wildland fire issues. Such a strategy should identify the available long-term options and associated funding for reducing excess vegetation and responding to wildland fires if the agencies and the Congress are to make informed decisions about an effective and affordable long-term approach for addressing problems that have been decades in the making. We first recommended in 1999 such a strategy be developed to address the problem of excess fuels and their potential to increase the severity of wildland fires and cost of suppression efforts.[7] By 2005, the agencies had yet to develop such a strategy, and we reiterated the need for a cohesive strategy and broadened our recommendation's focus to better address the interrelated nature of fuel reduction efforts and wildland fire response. The agencies said they would be unable to develop a cohesive strategy until they have completed certain key tasks. We therefore recommended that the agencies develop a tactical plan outlining these tasks and the time frames needed for completing each task and a cohesive strategy. These tasks include (1) finishing data systems that are needed to identify the extent, severity, and location of wildland fire threats in our national forests and rangelands; (2) updating local fire management plans to better specify the actions needed to effectively address these threats; and (3) assessing the cost-effectiveness and affordability of options for reducing fuels and responding to wildland fire problems.

First, federal firefighting agencies have made progress in developing a system to help them better identify and set priorities for lands needing treatment to reduce accumulated fuels. Many past studies have identified fuel reduction as important for containing wildland fire costs because accumulated fuels can contribute to more-severe and more costly fires. The agencies are developing a geospatial data and modeling system, called LANDFIRE, intended to produce consistent and comprehensive maps and data describing vegetation, wildland fuels, and fire regimes across the United States.[8] The agencies will be able to use this information to help identify fuel accumulations and fire hazards across the nation, help set nationwide priorities for fuel-reduction projects, and assist in determining an appropriate response when wildland fires do occur. LANDFIRE data are nearly complete for most of the western United States, with data for the remainder of the country scheduled to be completed in 2009. The agencies, however, have not yet finalized their plan for ensuring that collected data are routinely updated to reflect changes to fuels, including those from landscape-altering events, such as hurricanes, disease, or wildland fires themselves. The agencies expect to submit a plan to the Wildland Fire Leadership Council for approval later this month.

Second, we reported in 2006 that 95 percent of the agencies' individual land management units had completed fire management plans in accordance with agency direction issued in 2001.[9] As of January 2007, however, the agencies did not require regular updates to ensure that new data (from LANDFIRE, for example) were incorporated into the plans. In addition, in the wake of two court decisions—each holding that the Forest Service was required to prepare an environmental assessment or environmental impact statement under the National Environmental Policy Act (NEPA)[10] to accompany the relevant fire management plan—the Forest Service decided to withdraw the two plans instead of completing them. It is unclear whether the agency would withdraw other fire management plans successfully challenged under NEPA; nor is it clear whether or to what extent such agency decisions could undermine the interagency policy directing that every burnable acre have a fire management plan. Without such plans, however, current agency policy does not allow use of the entire range of wildland fire response strategies, including less aggressive, and potentially less costly, strategies. Moreover, in examining 17 fire management plans, a May 2007 review of large wildland fires managed by the Forest Service in 2006 identified several shortcomings, including that most of the plans examined did not contain current information on fuel conditions, many did not provide sufficient guidance on selecting firefighting strategies, and only 1 discussed issues related to suppression costs.[11]

Third, over the past several years, the agencies have been developing a Fire Program Analysis (FPA) system, which was proposed and funded to help the agencies

- determine national budget needs by analyzing budget alternatives at the local level—using a common, interagency process for fire management planning and budgeting—and aggregating the results;
- determine the relative costs and benefits for the full scope of fire management activities, including potential trade-offs among investments in fuel reduction, fire preparedness, and fire suppression activities; and
- identify, for a given budget level, the most cost-effective mix of personnel and equipment to carry out these activities.

We have said for several years—and the agencies have concurred—that FPA is critical to helping the agencies contain wildland fire costs and plan and budget effectively. Recent design modifications to the system, however, raise questions about the agencies' ability to fully achieve key FPA goals. A midcourse review of the developing system resulted in the Wildland Fire Leadership Council's approving in December 2006 modifications to the system's design. FPA and senior Forest Service and Interior officials told us they believed the modifications would allow the agencies to meet the key goals. The officials said they expected to have a prototype developed for the council's review in June 2007 and to substantially complete the system by June 2008. We have yet to systematically review the modifications, but after reviewing agency reports on the modifications and interviewing knowledgeable officials, we have concerns that the modifications may not allow the agencies to meet FPA's key goals. For example, under the redesigned system, local land managers will use a different method to analyze and select various budget alternatives, and it is unclear whether this method will identify the most cost-effective allocation of resources. In addition, it is unclear how the budget alternatives for local units will be meaningfully aggregated on a nationwide basis, a key FPA goal.

Although the agencies have made progress on these three primary tasks, as of April 2007, they had yet to complete a joint tactical plan outlining the critical steps, together with related time frames, that the agencies would take to complete a cohesive strategy, as we recommended in our 2005 report. We continue to believe that, until a cohesive strategy can be developed, it is essential that the agencies create a tactical plan for developing this strategy, so Congress understands the steps and time frames involved in completing the strategy.

Lack of Clear Goals or a Strategy Hinders Federal Agencies' Efforts to Contain Wildland Fire Costs

As we testified before the Senate Committee on Energy and Natural Resources in January 2007, the steps the Forest Service and Interior agencies have taken to date to contain wildland fire costs lack several key elements fundamental to sound program management, such as clearly defining cost-containment goals, developing a strategy for achieving those goals, and measuring progress toward achieving them. First, the agencies have not clearly articulated the goals of their cost-containment efforts. For cost-containment efforts to be effective, the agencies need to integrate cost-containment goals with the other goals of the wildland fire program—such as protecting life, property, and resources. For example, the agencies have established the goal of suppressing wildland fires at minimum cost, considering firefighter and public safety and values being protected, but they have not defined criteria by which these often-competing objectives are to be weighed. Second, although the agencies are undertaking a variety of steps designed to help contain wildland fire costs, the agencies have not developed a clear plan for how these efforts fit together or the extent to which they will assist in containing costs. Finally, the agencies are developing a statistical model of fire suppression costs that they plan to use to identify when the cost for an individual fire may have been excessive. The model compares a fire's cost to the costs of suppressing previous fires with similar characteristics. However, such comparisons with previous fires' costs may not fully consider the potential for managers to select less aggressive—and potentially less costly—suppression strategies. In addition, the model is still under development and may take a number of years to fully refine. Without clear program goals and objectives, and corresponding performance measures to evaluate progress, the agencies lack the tools to be able to determine the effectiveness of their cost-containment efforts. Our forthcoming report on federal agencies' efforts to contain wildland fire costs includes more-detailed findings and recommendations to the agencies to improve the management of their cost-containment efforts; this report is expected to be released at a hearing before the Senate Committee on Energy and Natural Resources scheduled for June 26, 2007.

Conclusions

Complex conditions have contributed to increasing wildland fire severity. These conditions have been decades in the making, and will take decades to resolve. The agencies must develop an effective and affordable strategy for addressing these conditions in light of the large federal deficit and the long-term fiscal challenges facing our nation. To make informed decisions about an effective and affordable long-term approach to addressing wildland fire problems, the agencies need to develop a cohesive strategy that identifies the available long-term options and associated funding for reducing excess vegetation and responding to wildland fires. Because the agencies cannot develop such a strategy until they complete certain key tasks, we continue to believe that in the interim the agencies must create a tactical plan for developing this strategy so that Congress can monitor the agencies' progress. While the agencies continue to work toward developing a cohesive strategy, they have initiated a number of efforts intended to contain wildland fire costs, but the agencies cannot demonstrate the effectiveness of these cost containment efforts, in part because the agencies have no clearly defined cost-containment goals and objectives. Without clear goals, the agencies cannot develop consistent standards by which to measure their performance. Further, without these goals and objectives, federal land and fire managers in the field are more likely to select strategies and tactics that favor suppressing fires quickly over those that seek to balance the benefits of protecting the resources at risk and the costs of protecting them. Perhaps most important, without a clear vision of what they are trying to achieve and a systematic approach for achieving it, the agencies—and Congress and the American people—have little assurance that their cost-containment efforts will lead to substantial improvement. Moreover, because cost-containment goals should be considered in relation to other wildland fire program goals—such as protecting life, resources, and property—the agencies must integrate cost-containment goals within the overall cohesive strategy for responding to wildland fires that we have consistently recommended.

REFERENCES

[1] The four agencies are the Bureau of Indian Affairs, Bureau of Land Management, Fish and Wildlife Service, and National Park Service.

[2] GAO, *Wildland Fire Management: Lack of a Cohesive Strategy Hinders Agencies' Cost-Containment Efforts*, GAO-07-427T (Washington, D.C.: Jan. 30 2007); *Wildland Fire Management: Update on Federal Agency Efforts to Develop a Cohesive Strategy to Address Wildland Fire Threats*, GAO-06-671R (Washington, D.C.: May 1, 2006); and *Wildland Fire Management Important Progress Has Been Made, but Challenges Remain to Completing a Cohesive Strategy*, GAO-05-147 (Washington, D.C.: Jan. 14, 2005).

[3] GAO, *Western National Forests: A Cohesive Strategy Is Needed to Address Catastrophic Wildfire Threats*, GAO/RCED-99-65 (Washington, D.C.: Apr. 2, 1999).

[4] GAO-05-147.

[5] The National Fire Plan is a joint interagency effort to respond to wildland fires. Its core comprises several strategic documents, including (1) a September 2000 report from the Secretaries of Agriculture and the Interior to the President in response to the wildland fires of 2000, (2) congressional direction accompanying substantial new appropriations in fiscal year 2001, and (3) several approved and draft strategies to implement all or parts of the plan.

[6] The Wildland Fire Leadership Council is composed of senior Agriculture and Interior officials, including the Agriculture Undersecretary for Natural Resources and

Environment; the Interior Assistant Secretary for Policy, Management, and Budget; the Interior Deputy Assistant Secretary for Business Management and Wildland Fire; and the heads of the five federal firefighting agencies. Other members include representatives of the Intertribal Timber Council, the National Association of State Foresters, and the Western Governors' Association.

[7] GAO/RCED-99-65.

[8] A fire regime generally classifies the role that wildland fire plays in a particular ecosystem on the basis of certain characteristics, such as the average number of years between fires and the typical severity of fire under historic conditions.

[9] Fire management plans are local plans prepared by individual agency management units (such as national forests or wildlife refuges) to define each unit's program to prepare for and manage fires; such plans are important for identifying the fuel reduction, preparedness, suppression, and rehabilitation actions needed at the local level to effectively address wildland fire threats.

[10] For major federal actions that significantly affect the quality of the human environment, the National Environmental Policy Act requires all federal agencies to analyze the environmental impact of the proposed action. 42 U.S.C. § 4332(2)(C).

[11] Independent Large Wildfire Cost Panel, chartered by the U.S. Secretary of Agriculture, *Towards a Collaborative Cost Management Strategy: 2006 U.S. Forest Service Large Wildfire Cost Review Recommendations* (Washington, D.C., May 15, 2007).

INDEX

A

B

C

D

E

R

S